Schola Ludens Europaea: papers dedicated to Roland Naul

Edited by Claude Scheuer and Dennis Dreiskämper

Schola Ludens Europaea: papers dedicated to Roland Naul

Meyer & Meyer Sport

British Library Cataloguing in Publication Data
A catalogue record for this book is available from the British Library

Schola Ludens Europaea
Maidenhead: Meyer & Meyer Sport (UK) Ltd., 2019
ISBN: 978-1-78255-136-2

Aachen, Auckland, Beirut, Cairo, Cape Town, Dubai, Hägendorf, Hong Kong, Indianapolis, Manila, New Delhi, Singapore, Sydney, Tehran, Vienna

Member of the World Sport Publishers' Association (WSPA), www.w-s-p-a.org

Printed by Bookwire
ISBN: 978-1-78255-136-2
E-Mail: info@m-m-sports.com
www.m-m-sports.com

Contents

PART ONE:

ROLAND NAUL:
SCHOLA LUDENS EUROPAEA

1 Foreword

Claude Scheuer and Dennis Dreiskämper

There are very good reasons to honour Roland Naul with a commemorative for his great achievements in his professional life. It bears the title *Schola Ludens Europaea* and thus treats the focal points of Roland Naul's scientific and political impact on a European and international level. Authors who have been in professional and personal contact with Roland Naul in recent years have spontaneously agreed to contribute to this volume. In the first part dedicated to Roland Naul as a Schola Ludens Europaea, Jacob Kornbeck appreciates the merits of Roland Naul as a sport social pedagogue when it comes to the increasing importance of health-enhancing physical activity (HEPA) on the EU sport policy agenda. This text is based on a keynote speech Jacob Kornbeck delivered at Charles University in Prague on January 31st 2017 as part of the mini-symposium organized in connection with the conferral of a title of doctor of philosophical sciences, honoris causa, on Roland Naul. Deanna Binder then attempts to bring to life the odyssey of Roland Naul in pursuit of a pedagogy for Olympic education. This section concludes with a compilation of Roland Naul's football research in the last 40 years, presented by David Niehues.

In part two, 'General education and sport education', Niels Rosendal Jensen and Søren Nagbøl analyze the traditions of sports and social work based on the description of the morphology of a city with the example of Copenhagen, with the aim of contributing elements to a rethinking of the relationship between sports and social pedagogy. Nils Neuber continues with the presentation of the status quo and challenges for physical activity and sports in German all-day schools.

Part three analyzes 'Old and new challenges for the Olympic Games movement'. First, March L. Krotee reflects on sociological perspectives on the Olympic Games from Olympism by examining the development of the International Olympic Committee in an attempt to elucidate some of the problems and issues concerning the study of this sociologic reference group. In a second contribution, Michael Krüger attempts to explain what Olympism really means from a historical perspective and discusses three perspectives of Olympic education as crucial elements of modern education as a whole.

In part four, the focus lies on 'Physical activity and health: HEPA policy and development in the European Union'. In a first contribution, Claude Scheuer discusses the challenge of HEPA for physical education in Europe. Dennis Dreiskämper and Till Utesch give an example of an intervention study to reduce overweight and to boost physical abilities in EU countries with the case of the GKGK/HCSC (Healthy Children in Sound Communities) concept. As the promotion of HEPA is one of the priorities of the Erasmus+ programme of the European Commission, a third contribution in this section gives an overview of Erasmus+-funded European projects which the European Physical Education Association (EUPEA) is involved in (Claude Scheuer, Rose-Marie Repond and Martin Holzweg). Jacob Kornbeck closes this section with a confrontation of *structures* versus *activities* in public policies for the promotion of sport and physical activity (HEPA) by arguing in favour of a conceptual framework relying on the notion 'le geste sportif'.

Part five turns its focus on 'Sport policy in the local context', with a contribution analyzing the weight status of children with a Turkish or Syrian background in Germany and the Netherlands and arguing for early interventions based on daily physical activity and weekly lessons on healthy food at school (Monique L'Hoir, Sunay Nuraydin and Dennis Dreiskämper).

In the appendix, Jacob Kornbeck and Dennis Dreiskämper compile the complete work and publications of Roland Naul from 1972 to 2018.

On the one hand, with all these contributions, the present commemorative pursues the goal of paying tribute to Roland Naul's outstanding contribution to the subject areas covered in this book. On the other hand, this work should give Roland Naul a pleasure. It is to honour his outstanding merits in his scientific field and to express his colleagues' gratitude for the pleasing and lasting cooperation based on respect, which will remain in memory.

We congratulate Roland Naul on his 70th birthday and are pleased that he continues to stand amongst the international scientific community in physical education and school sport with his experience and knowledge. We sincerely wish him the very best for his professional and especially private future!

2 Greeting by the director of the Institute of Sport and Exercise Sciences, WWU Münster

Nils Neuber

To many national and international colleagues, Roland Naul is known for his works at the University of Duisburg-Essen. For three decades – from 1980 until 2013 – he functioned as a professor for sports science and sports pedagogy in Essen. Actually, Roland Naul is a 'Münster child' though. From 1968 to 1974 he studied physical education, pedagogy, psychology, sociology, sports medicine and Russian at the University of Münster. In 1978 he earned his PhD with a work on A-level schools and worked as a research assistant with Professor Herwig Blankertz until 1980. Afterwards he transferred to the University of Duisburg-Essen as a regular professor, where he established his reputation in the areas of school sports, child and youth sports, Olympic education, sports history and not least international networking.

In 2013, when Roland Naul returned to the University of Münster as a senior professor, he could look back on a rich wealth of experience, which we gladly took up. His workspace for 'European Studies in Physical Education and Youth Sport' benefits the research and lecture offers at the Institute of Sports and Exercise Sciences profitably. With currently 1,600 students, 80 colleagues and 8 professorships, our institute belongs to the biggest teacher training sites in Europe. Furthermore, by having new English-speaking courses of study like BSc Human Movement in Sports and Exercise and MSc Sports, Exercise and Human Performance, we are interesting for students from all over the world. Not least, we maintain fruitful cooperation with partner universities from Europe and outside of it.

One of Roland Naul's most valuable credits is to have increased the international visibility of the Institute of Sports and Exercise Sciences significantly. Besides his numerous personal contacts in Europe and all over the world; he initiated partnerships with the universities Arnhem en Nijmegen (Netherlands), Prague (Czech Republic), Tongji (Shanghai) and Tsukuba (Japan) amongst others. His most outstanding events while working in Münster are the conference of the Conseil Europe'en des Recherches en Education Physique et Sportive (CEREPS) in the NRW-house in Brussels in June 2017, his contribution to the German-Japanese symposium in Münster in March 2018 as well as the EuPEO project focusing on European school sport monitoring, which kicked off in Lisbon in January 2018. Beyond debate, the international highlight was the honorary doctorate, which the Karls University awarded Roland Naul in Prague in January 2017 for his long-term credits in international collaboration.

We congratulate Roland Naul on his 70th birthday and are glad that he continues to stand beside us with his experiences and expert knowledge. We sincerely wish him all the best for his professional and private future!

3 Greeting by the president of the Willibald Gebhardt Institute e.V.

Maike Tiejens

For me, Roland Naul is inseparable from the Willibald Gebhardt Institute e.V. (WGI) and the many activities that result from it. The WGI was founded in 1992 in Essen and is a registered non-profit association whose purpose is to promote scientific, social science, psychology and educational research in the field of Olympic sport, competitive sports, school and club sports and health sports. In doing so, the WGI has been tasked with ensuring the practical application and exchange of scientific knowledge, attracting and supporting research projects, and collaborating with other research institutions, societies and associations both at home and abroad. Around 25 years later, based in Münster today, about 20 sports and science institutions (regional, national, international) belong to the WGI and a greater number of private individuals as scientists from Germany, Europe and Asia are member of the WGI.

The members of the WGI, and especially Roland Naul, have carried out many national and international projects, including the Healthy Children in Healthy Communities project, which is now being carried out in various countries, book projects on school and club Olympic education; lifestyle development in Europe and Talent development in football.

The success story of the WGI is inextricably linked with the name Roland Naul, as a founding member, as a long-time chairman, board member and today as vice president of the WGI. Roland Naul has shaped the development of the WGI like nobody else and made it what it is today. Roland Naul has worked tirelessly to represent the WGI at national and international congresses and has established contacts with national and international sports associations and institutions. This is also reflected in the fact that he is in demand worldwide as a speaker.

Hence, it is not surprising that my first personal meeting with Roland Naul was during an AIESEP (Association Internationale des Écoles Supérieures d'Éducation Physique) meeting in Jyvaskyla, Finland. I can still remember this conversation very well, and in retrospect it becomes clear that he wants to convey his credo of an international sport science for the benefit of sport to young scientists. From then on, we met again and again at international conferences, amongst others at the German-Japanese Symposium, which was and is regularly held with the participation of the WGI. Out of the many conversations, joint projects that linked his topics and mine arose. Today we work together in the presidency of the WGI, and I am grateful that he supports me with words and deeds.

Over the years, I have seen Roland Naul as a committed networker in sports policy, as a footballer at heart, as an enthusiastic representative of the Olympic Movement, and an internationally renowned scientist who has set up the WGI for the future.

On behalf of the presidium of the WGI, I congratulate Roland Naul very warmly on his 70th birthday, and I hope that he will remain involved with the WGI with advice and action and will remain esteemed.

4 Adducing HEPA to the EU sport policy agenda: Roland Naul as a sport social pedagogue and a modern *schola ludens* in the spirit of Comenius[1]

Jacob Kornbeck[2]

I am pleased and honoured to be able to deliver the introductory remarks to this session and, thereby, to honour my friend Roland Naul and his contribution, not only to the advancement of academic research and teaching, but indeed also to EU policy development over the last 13 years. As I am an EU official, please bear in mind that these are personal comments, not one made in the name of the EU. 'Adducing HEPA to the EU sport policy agenda' is the title, while the subtitle reads: 'Roland Naul as a sport social pedagogue and a modern *schola ludens* in the spirit of Comenius.'

> *The virtues are learned by constantly doing what is right [. . .] it is by learning that we find out what we ought to learn, and by acting that we learn to act as we should. So then, as boys easily learn to walk by walking, to talk by talking,*

1 *Reworked text of a keynote speech held at Charles University Prague, 30 January 2017, as part of the mini-symposium organised in connection with the conferral of a title of doctor of philosophical sciences, honoris causa, to Roland Naul. Greeting formulas used in the speech were removed as part of the editing process.*

2 *Previously policy officer in the European Commission, Sport Unit (2001–14). Currently policy officer in the European Commission, Youth Unit. Disclaimer: Remarks made strictly in a private capacity. No official EU positions expressed.*

and to write by writing, in the same way we will learn obedience by obeying, abstinence by abstaining, truth by speaking the truth, and constancy by being constant. But it is necessary that the child be helped by advice and example at the same time.

Comenius, Didactica Magna, Ch. XXIII, p. 367, cit. Piaget, 1997, p. 182[3]

Today you are united in Prague for the mini-symposium organized in connection with the conferral of an honorary doctorate upon Roland Naul, professor emeritus of the University of Essen, president of the Willibald Gebhardt Institute, previously of Essen but recently transferred to Münster, and currently senior professor of European Studies in Physical Education and Youth Sport, Institute of Sport and Exercise Sciences, University of Münster, Germany.

Roland is a colleague and a friend whom I have known since 2004. As policy officer in the European Commission, Sport Unit (2001–2014), I was in charge, inter alia, of the health-enhancing physical activity (HEPA) file which transformed the EU's approach to sport policy.

My policy development work was greatly inspired by the research and the projects of Roland Naul and his colleagues, and I learnt to respect Roland as an outstanding academic who can also talk to practitioners. While I regret that I cannot be with you today in Prague, I am sure that the Great Hall of the Carolinum is a splendid setting for the ceremony, just like the FTVS faculty (Faculty of Physical Education and Sport) premises will be for the mini-symposium, both of which are befitting ways to honour my respected friend Roland.

What I would like to demonstrate in this speech of mine is that Roland Naul, besides being an eminent researcher and academic teacher, is also an intellectual whose sharp mind has proven invaluable in informing policy development, be it at the local, regional, national or European level. Roland's capacity to draw upon the best available research-generated knowledge and infuse it into proposals for action makes him stand out.

3 *Quotation not included in the initial text.*

For academia to share its knowledge with those who develop the policies of tomorrow is not only a noble task, but also a daunting one, as it involves translating knowledge validated within academia into information that can be used by busy decision-makers – yet without over-simplifying, without distorting facts and while remaining faithful to the standards of academic objectivity.

But what about the title of my speech?

But what about the title of my speech? 'Adducing HEPA to the EU sport policy agenda' seems clear enough: the HEPA concept transformed what had essentially been policies directed towards organized sport, in particular regarding the regulation of the economic activities of sports governing bodies and their relations with athletes and third parties. HEPA made EU sport policy much richer, and the Lisbon treaty affirmed this approach and elevated it to the status of EU 'soft law' (*Kornbeck, 2006a; 2015*).

'Roland Naul as a sport social pedagogue', by contrast, is a phrase that may require further explanations: it blends knowledge and skills related to sport with those pertaining to social pedagogy which, in turn, is concerned with education as well as with social issues. Sport social pedagogy is an effective way to characterise the area in which Roland is a European trailblazer and an expert. Social pedagogy is familiar to colleagues in parts of the continent and in parts of the Nordic countries, while the United Kingdom has just started learning about it (see, for example, Kornbeck, 2013). Social pedagogy draws on a holistic concept of learning and one that allows addressing social issues by educational means. It is not unlike the dictum of the great Czech educationist Comenius who wanted his Great Didactic to represent 'the whole art of teaching all things to all men' [*sic*] *(cit. Piaget, 1997, p. 175*). But 'all things' cannot be limited to the cognitive ambit, as indeed Comenius was convinced of the usefulness of play.

> *But he who shows the way is not necessarily an adult. In a curious passage of the Methodus linguarum novissima, quoted by P. Bovet [1943, p. 10], Comenius lays stress on imitation and group games, bringing his systematic mind to outlining the seven characteristic factors of such games. He appears, in this connection, to have recognized the role of the social relationship set up among players of*

games, as well as the role of competition and the rules imposed upon players by the game.[4] Piaget, 1997, p. 182

Social pedagogy has a great potential to enrich sport science and sport pedagogy, though social pedagogy literature is rarely quoted in sport pedagogy books.[5] Now, I would argue that Roland Naul's contribution to the process whereby the HEPA concept was embedded into EU sport policy was based on an understanding of the relevant issues which can be characterised as sport social pedagogy. Having made these lateral connections makes Roland Naul stand out once more.

Now, having started off with Comenius, let us add some West German pop lyrics from the relevant decades, as a tribute to Roland Naul.[6]

The 1970s

Wir fahr'n fahr'n fahr'n auf der Autobahn
Vor uns liegt ein weites Tal
Die Sonne scheint mit Glitzerstrahl
Die Fahrbahn ist ein graues Band
Weiße Streifen, grüner Rand
Jetzt schalten wir ja das Radio an
Aus dem Lautsprecher klingt es dann:
Wir fah'rn auf der Autobahn. . .[7]

Kraftwerk, Autobahn, 1974[8]

Looking at Roland Naul's publications of the last 45 years, athlete welfare and social interactions in sport were at the forefront of Roland's concerns in two papers co-published

4 *Quotation not included in the initial text.*

5 *For a discussion see from the perspective of German and Austrian sport pedagogy. See Kornbeck (2006a).*

6 *Text not included in the initial text.*

7 *Quotation not included in the initial text.*

8 *See http://www.songtextemania.com/kraftwerk_songtexte.html*

with his colleague Voigt in 1972 and 1974 (Naul & Voigt, 1972, 1974). Roland was then 23 and 25 years old, respectively, and had yet to earn his masters' degree – I was then four and six years old. This reference is not without interest as I see Roland as a sort of spiritual father. What Turnvater Jahn was to nineteenth-century German gymnastics, HEPA-Vater Naul has been to me. Note that the 1974 paper was published in a mainstream social science journal, *Soziale Welt.* Getting sport content out into the mainstream can be a challenge, yet it is very rewarding when you succeed. This paper provides a view on sport which would fit well into a social work publication, but since Roland Naul's thinking is grounded to a large extent in education in the German tradition of *historische und systematische Pädagogik*, the most appropriate term is *pedagogy* or *social pedagogy.* Roland Naul and his colleagues did research related to the curricular reforms of the early 1970s, where different streams of secondary education, including sports studies, were defined as contributing towards the attainment of more global, overarching competences (see Heursen, Naul & Gruschka, 1978; Fingerle, Naul & Schenk, 1978). Today this may seem almost trivial; in those days, it marked a watershed.

What, then, if the conception of education in its widest sense is applied to sport? Let's call this sport social pedagogy: no one has claimed this term before – we'll be the first to do so. And the process continued during the ensuing decades.

> *Olympic education includes a number of pedagogical concepts. Some of these share common historical and pedagogical concepts; others take this in mind to focus on selected Olympic values out of the Olympic Charter. Some concepts do not merely seek to promote a sporting lifestyle by way of practical lessons, but prefer to teach historical and current national items of the Olympic Games movement with a cognitive approach, while others also encourage certain social attitudes and behaviours in different subjects and lessons of the school curriculum and in extra-curricular physical activities.* Naul, 2008, 102[9]

9 *Quotation not included in the initial text.*

The 1980s and the 1990s

Entschuldigen Sie, ist das der Sonderzug nach Pankow
Ich muß mal eben dahin, mal eben nach Ost-Berlin
Ich muß da was klären, mit eurem Oberindianer
Ich bin ein Jodeltalent, und ich will da spielen mit ,ner Band

[. . .]
Ich weiß genau, ich habe furchtbar viele Freunde
In der DDR und stündlich werden es mehr
Och, Erich ey, bist Du denn wirklich so ein sturer Schrat
Warum lässt Du mich nicht singen im Arbeiter- und Bauernstaat?
Ist das der Sonderzug nach Pankow?[10]

Udo Lindenberg, Sonderzug nach Pankow, 1983[11]

In the same vein, in 1986 – by then I was 18 years old – Roland Naul co-authored a paper, with Fischer as the lead author, on pupils' transition to university studies and/or working life, which was published in a mainstream education journal, the *respected Zeitschrift für Pädagogik* (Fischer, Gruschka, Meyer, Naul & Schenk, 1986). Again, this was writing for a wider academic audience, leaving the narrow confines of sport science and PE teacher training to share their knowledge with wider communities. In the following year – I was 19, had started my university studies and the Iron Curtain was still in place – Roland Naul (without co-authors) published a paper in an edited book published by the very same mainstream education journal (as a special supplement issue: *Beiheft der Zeitschrift für Pädagogik):* 'Sport education as part of a new general education' *(Naul, 1987).* The semantic similarity with US sport education, a concept which was coined by Daryl Siedentop (1994) seven years later, is obvious. Intriguingly, Roland has here made a contribution to curriculum debates located outside his own academic community. With his focus on the skills taught through PE and sport education, he shows himself as a sport

10 Last text not included in the initial text.

11 See http://www.songtexte.com/songtext/udo-lindenberg/sonderzug-nach-pankow-1bda7514.html

social pedagogue. One year later, in 1988 – I was then 20 and the Berlin Wall and the Iron Curtain had only few months left – Roland Naul published a paper in the *International Journal of Physical Education* entitled 'Learning about sports: a cognitive approach and its evaluation in the German Grammar School physical education curriculum' (Naul, 1988). While noting that 'cognitive learning about sports is still a new challenge for all' (ibid., 21) Naul remained optimistic that[12]

> *curriculum revision can improve the interrelations between all types of exercises and all types of scientific subjects of sports because it provides the basis for interrelations in each grade and not 'only sometimes' [. . .]. Apart from all improvements in curriculum structure, however, and in accordance with the interests of the students a new conception of teachers' training would be necessary. Otherwise there will be no future of cognitive learning about sports in the school curriculum.*
> Naul, 1988, 22[13]

Many of the ideas expressed there can still be found in his seminal, recent textbook on Olympic education (Naul, 2008). And indeed, a book chapter printed in 1998 dealt with 'Olympic ideals and Olympic education of youth' (Naul, 1998). To this day, Roland Naul remains a reference figure for Olympic education, with numerous invitations to Olympic congresses and special seminars organized by National Olympic Academies, such as the one here in Prague, as well as the IAO (International Olympic Academy) in Olympia, Greece.

> *Because the school is only one setting among many in which young people can live and learn, exercise and practice sports, Olympic education is not limited to scholastic teaching. Important aspects of learning and experience are addressed to young people in their various settings outside school as well as in their various leisure activities, e.g. with multimedia tools. Very often, social virtues and moral values – good ones and bad ones – are conveyed to boys and girls far more by informal learning than by formal teaching. Therefore, as far as Olympic education*

12 *Quotation not included in the initial text.*

13 *Quotation not included in the initial text.*

is concerned, both informal learning and formal teaching are core elements that go beyond the school as one setting of life and go beyond scholastic lessons as a special part of learning. Naul, 2008, 164[14]

Roland never lost his interest in the services rendered by PE to the curriculum more generally, and he kept a keen eye on the various skills taught in PE. That was also *en vogue* in EU policy circles as 1992 was the year when the ambitious Single Market programme neared its completion (Pépin, 2007). I was 24, studying in Germany and gradually turning my attention towards German *Pädagogik and Sozialpädagogik*. I did not yet know Roland Naul, yet I was interested in many of the issues he was then working on, including our neighbours beyond the former Iron Curtain. My first visit to Prague took place in March 1990: an unforgettable experience.

The 2000s and the new millennium

L'enfer du Nord Paris-Roubaix
La Cote d'Azur et Saint Tropez
Les Alpes et les Pyrénnées
Dernière étape Champs-Elysées
Galibier et Tourmalet
En danseuse jusqu'au sommet
Pédaler en grand braquet
Sprint final a l'arrivée
Crevaison sur les paves
Le vélo vite réparé
Le peloton est regroupe
Camarades et amitié[15]

Kraftwerk, Tour de France, 2005[16]

14 *Quotation not included in the initial text.*

15 *Quotation not included in the initial text.*

16 *See http://www.songtextemania.com/tour_de_france_songtext_kraftwerk.html*

For once, the lyrics are even sport-related. And far from being narrowly confined to the 'agonistic' aspects of competitive sports, they refer to the communities that emerge when people are jointly engaged in games.[17]

Roland Naul never dropped his initial interest in the conflicts and tensions of the sporting world. The topic of the early 1970s was taken up again after the turn of the millennium, now by looking more at the mental strain involved in elite sport – an activity which can take a high toll – in a co-authored 2006 book chapter addressing the psycho-mental personality traits of young elite footballers (Naul, Richter, Hoffmann, Altenbeck, Brüggemann & Fischer, 2006). Implicitly, this could also be interpreted as a contribution to the dual careers debate which had taken off, at EU level, in 2004, the European Year of Education through Sport. I had joined the European Commission's Sport Unit in 2001 and had met Roland in 2004 when Wolf-Dietrich Brettschneider and he had won a contract to conduct, for the commission, the first-ever structured review covering all (then) 25 EU Member States and addressing the challenges posed by decreasing daily physical activity and increasing overweight and obesity levels. The Paderborn-Essen study of 2004 (Brettschneider & Naul, 2005) became the bedrock and bottom line for all subsequent EU policy development in this field.

That was about sport-for-all, not elite sport. Yet 2004 was also the year when other commission-sponsored studies focused on how elite athletes cope with the challenges of developing half-way normal lives (Eureval-C3E, 2005, p. 39, 'Figure 4: A small team for such a big project'). The 2006 book's chapter fits in very well there too. In 2009 – I was then 41 – Richter & Naul provided an insightful study into elite footballers' fear of not being up to the mark (Richter & Naul, 2009). The German language is famous for having coined the word *Angst*. Richter & Naul phrase athletes' number one stressor using the equally Teutonic neologism *Versagensangst*, thereby providing a conceptual framework which would be excellent for addressing, for example, the dilemmas of Belgian footballer Jean-Marc Bosman. In 1995, he had won a landmark EU court case, liberating his colleagues from rather feudal restrictions on their free movement as workers, yet he paid a very high price for this courage, ending up living from social welfare (Slater, 2015).

---

17 *Text not included in the initial text.*

Researchers like Richter & Naul can help decision makers and the general public better appreciate the precarious nature of sporting careers, looking behind the shiny façade and realising that the seemingly strong individuals may be rather vulnerable indeed. Combining insights into learning processes and social risks is a classic of social pedagogy. Applying that to sport is sport social pedagogy – although it still seems as if sport pedagogy and sport pedagogy actors (be it in academia or in front-line practice) do not know much about each other (Kornbeck, 2006b).

After 2004, a process took off whereby the EU started using the evidence from the Brettschneider-Naul study and linking it with the HEPA concept which had until then been promoted by the WHO Europe Region. The concrete Brettschneider-Naul policy recommendations are numbered at the end of the study. They are particularly concerned with the fostering of local multi-actor networks addressing insufficient daily physical activity and linking these local networks with similar ones in other EU countries. This became a reference for EU projects in the field of sport: what is known under the Erasmus+ Programme as 'collaborative partnerships' (European Commission, 2017, pp. 227–235). In the coming years Roland Naul and his colleagues formed the scientific back office of a range of such projects, some of which were successful in attracting EU funding. They shared their experiences in a number of publications (See, for example, Brettschneider, Hoffmann, Naul & Steinzen, 2009; Naul, Schmelt, Dreiskämper et al., 2012).

In 2013, in a paper published in the *German Journal of Sports Medicine*, Roland Naul would share his ideas on what makes some networks work, while others not (Naul, 2013). I was then 45, and the EU had given itself the Lisbon treaty. Sport had finally been taken into the treaty, as the Lisbon treaty had seen the education article amended to also include sport in what is now Article 165 of the Treaty on the Functioning of the European Union (TFEU). Continuous, multiannual funding for sports projects was finally made available through Erasmus+: that this money would not just be for sport as such but also for sport-for-all and HEPA, had much to do with the policy development work which had taken place between 2004 and 2009, the year when the new treaty entered into force. Before Erasmus+ came into being, so-called preparatory actions allowed testing the pertinence of various topics and models, each year on a different basis. Naul and his colleagues made a solid contribution in this connection, with their local intervention project 'healthy children in sound communities' *(see, for example, Naul, Schmelt & Dreiskämper, et al., 2012).*[18] This is all sport social pedagogy, yet adapted for busy decision makers.

For obvious reasons I cannot share too many details of this important work with you: I simply don't have the time to do so, and indeed, another speaker will deal with this later. But I would like to emphasise that Roland's contribution took place at many different levels: as an academic, as a consultant, as a networker and as an expert on various expert groups organized by the commission as requested by member states. Roland fitted wonderfully into all of these roles: having him on board was always a gain. His ability to bring the most diverse actors together is unparalleled. On one occasion I travelled to Velen, a semi-rural town close to Borken in Roland's native region close to the Dutch-German border, for a conference launching one of Roland's EU-co-funded HEPA projects. Funding did not then come from one of the sport policy budgets but rather from Interreg, the programme allowing partners in border regions to run innovative projects with partners across the border. I was there as the local mayor, along with representatives of schools and sport clubs, the local baker (there was also a nutrition aspect in that project) and the local radio network. This was local multi-actor cooperation at its best, because Roland Naul was the brain behind it all, with an EU vision.

Roland Naul, the European sport social pedagogue

Trans Europa Express
Trans Europa Express
Trans Europa Express
Trans Europa Express
Rendez-vous auf den Champs Elysées
Verlass Paris am Morgen mit dem TEE
In Wien sitzen wir im Nachtcafé

18 *See also the PowerPoint presentation published on the website of the European Commission, Directorate General for Public Health: http://ec.europa.eu/health//sites/health/files/nutrition_physical_activity/docs/ev20111006_co04_en.pdf.*

Direkt Verbindung TEE
Wir laufen 'rein in Düsseldorf City
Und treffen Iggy Pop und David Bowie

Kraftwerk, 1977[19]

So here we are now, me in Brussels and you in Prague. The year is 2017, 45 years after the first Naul paper quoted in my speech. I am now 48, and my spiritual father Roland Naul has retired from Essen University, only to join Münster University as the head of a much sough-after EU project team. He remains a reference in the field, an outstanding academic and networker. At times when the European project is suffering some damage caused by the Brexit vote as well as by populism in some member states, we need people like Roland who can show the added value of being in a strong union. Projects between partners in different EU countries are an ongoing process and a never-ending story, always being 'on a learning curve' as today's expression goes – or, to borrow from Comenius: 'it is by learning that we find out what we ought to learn' *(Piaget, 1997, p. 182)*, including when it comes to policy development.

Roland is able to do this by providing the scientific evidence needed, by advising policymakers and by leading innovative projects. He is able to work with people from all walks of life, and he is a sport social pedagogue. His honorary degree is well deserved. I would like to congratulate you, Roland, on your achievements and the recognition you are receiving in Prague right now, and I will conclude by wishing all of you a successful and thought-provoking conference. I know that Roland's Czech friends, including a member of the board of the journal *Kinanthropologica*, have always been dear to him, and it seems appropriate that he should receive this distinction from Charles University. I believe that Comenius would have approved of your decision, had he known about Roland Naul's work, as a modern *schola ludens*, on physical activity, games and sport, including in relation to recent developments in the German *Ganztagsschule* – first in Naul (2005), then in numerous ensuing publications.

---

19 See http://www.songtextemania.com/transeuropa_express_songtext_kraftwerk.html

So congratulations to Roland Naul, our *schola ludens*, and to all colleagues reunited in Prague, a successful conference in the spirit of Comenius!

References

Bovet, R. (1943). *Jean Amos Comenius: un patriote cosmopolite.* Geneva: Port-Noir (Cit. Piaget, 1997)

Brettschneider, W. D., Hoffmann, D., Naul, R., & Steinzen, A. (2009). Von den Erkenntnissen aus dem EU-Bericht "Young people,s lifestyle and sedentariness" zur handlungsorientierten Umsetzung der Empfehlungen im Rahmen von zwei lokalen Präventionsstudien. In R. Naul, A. Krüger, & W. Schmidt (Eds.), *Kulturen des Jugendsports. Bildung, Erziehung und Gesundheit* (pp. 67–104). Aachen: Meyer & Meyer.

Brettschneider, W. D., & Naul, R. (2005). *Study on young people's lifestyles and sedentariness and the role of sport in the context of education and as a means of restoring the balance.* Retrieved from http://www.uni-muenster.de/imperia/md/content/sportwissenschaft/eu-studies/abstracts/noored_ja_sport_04_lotpaderborn.pdf

Eureval-C3E. (2005). *External evaluation (ex-post) of the European Year of Education through Sport EYES 2004.* Retrieved from http://ec.europa.eu/dgs/education_culture/more_info/evaluations/docs/sport/eyes_en.pdf

European Commission. (2017). *Erasmus+ Programme Guide. Version 2 (2017):* 20/01/2017 (2017). Retrieved from http://ec.europa.eu/programmes/erasmus-plus/sites/erasmusplus/files/files/resources/erasmus-plus-programme-guide_en.pdf

Fingerle, K., Naul, R., & Schenk, B. (1978). Schwerpunktbezogene Grundbildung in der Kollegschule: Problemstellungen und Konstruktionswege – Bilanz und Perspektiven. In B. Schenk & A. Kell (Eds.), *Grundbildung: Schwerpunktbezogene Vorbereitung auf Studium und Beruf in der Kollegschule.* Königstein/Ts.: Athenäum Verlag (Athenäum Taschenbücher; 3149), 255–284.

Fischer, B., Gruschka, A., Meyer, M., Naul, R., & Schenk, B. (1986). Schüler auf dem Weg zu Studium und Beruf. *Zeitschrift für Pädagogik, 32*, 557–577.

Heursen, G., Naul, R., & Gruschka, A. (1978). Die sozialberufliche Grundbildung – Konzept und Struktur. In: Schenk, B., & Kell, A. (Eds.), *Grundbildung: Schwerpunktbezogene Vorbereitung auf Studium und Beruf in der Kollegschule.* Königstein/Ts.: Athenäum Verlag (Athenäum Taschenbücher; 3149), 195–221.

Kornbeck, J. (2006a). Governance als Soft Law: Innovation oder Notwendigkeit im sportpolitischen Handeln der EU? In: Tokarski, W., Petry, K., & Jesse, B. (Eds.), *Sportpolitik. Theorie- und Praxisfelder von Governance im Sport.* Cologne: Sportverlag Strauß (Veröffentlichungen der Deutschen Sporthochschule Köln, vol. 15), 31–52.

Kornbeck, J. (2006b). Kennt die deutsche Sozialpädagogik die deutsche Sportpädagogik? Fachtheoretische Überlegungen zu einer nahe liegenden Partnerschaft. In A. Heimgartner & K. Lauermann (Eds), *Kultur in der Sozialen Arbeit. Festschrift für Univ.-Prof. Dr. Josef Scheipl.* Klagenfurt: Mohorjeva, 363-385.

Kornbeck, J. (2013). "Not a prophet" versus "foreign fads": Two scenarios for conceptualising social pedagogy dissemination. *Scottish Journal of Residential Child Care, 12*(2), 3–12.

Kornbeck, J. (2015). Lisbonisation without regulation: Engaging with sport policy to maximise its health impact? *International Sports Law Journal,* 15(1–2), 112–122.

Naul, R. (1987). Sporterziehung als Bestandteil einer neuen Allgemeinbildung. In H. Heid & H. G. Herrlitz (Eds.), *Allgemeinbildung.* Weinheim: Beltz (21. Beiheft der Zeitschrift für Pädagogik), 161-171.

Naul, R. (1988). Learning about sports: A cognitive approach and its evaluation in the German Grammar School physical education curriculum. *International Journal of Physical Education, 25,* 16–22.

Naul, R. (1998). Olympic ideals and Olympic education of youth. In R. Naul, K. Hardman, M. Pieron, & B. Skirstad (Eds.), *Physical activity and active lifestyle of children and youth.* Schorndorf: Hofmann, 29-48.

Naul, R. (2005). Bewegung, Spiel und Sport im offenen Ganztag. *Sportunterricht, 54,* 68–72.

Naul, R. (2008). *Olympic education.* Aachen: Meyer & Meyer.

Naul, R. (2013). Netzwerke – wie funktionieren sie effektiv? *Deutsche Zeitschrift für Sportmedizin, 64*(5), 143.

Naul, R., Richter, C., Hoffmann, D., Altenbeck, B., Brüggemann, D., & Fischer, B. (2006). Fußball-Fitness und psycho-mentale Persönlichkeitsmerkmale bei D-Junioren im Kader des Fußball- und Leichtathletik-Verbandes Westfalen (FLVW). In Weber, K. (Ed.), *Wissenschaftlicher Transfer für die Praxis der Sportspiele*. Cologne: Bundesinstitut für Sportwissenschaft, 132-143.

Naul, R., Schmelt, D., Dreiskaemper, D., et al. (2012). "Healthy children in sound communities" (HCSC/gkgk) – A Dutch-German community-based network project to counteract obesity and physical inactivity. *Family Practice, 29*(1), 110.

Naul, R., & Voigt, H. (1972). Aggressionsverhalten bei Volleyballspielern. *Die Leibeserziehung, 21*, 83–87.

Naul, R., & Voigt, H. (1974). Aggression und Sport. *Soziale Welt, 25*, 347–369.

Pépin, L. (2007). The history of EU cooperation in the field of education and training: How lifelong learning became a strategic objective. *European Journal of Education, 42*(1), 121–132.

Piaget, J. (1997). Jan Amos Comenius. In Z. Morsy (Ed.), *Thinkers on education. Vol. 1*. Paris & Oxford: UNESCO & IBH Publishing, 173-196.

Richter, C., & Naul, R. (2009). Versagensangst und Selbstvertrauen als psycho-mentale Merkmale in Relation zur Fußball-Fitness bei C- und D-Junioren des Fußball- und Leichtathletikverbandes Westfalen (FLVW). In U. Frick (Ed.), *Fußball in Schule und Verein* . Hamburg: Czwalina, 99-106.

Siedentop, D. (1994). *Sport education: Quality PE through positive sport experiences*. Champaign, IL: Human Kinetics.

Slater, M. (2015). Bosman ruling: 20 years on since ex-RFC Liege player's victory. *BBC Sport*, December 15, 2015. Retrieved from http://www.bbc.com/sport/0/football/35097223

5 Roland Naul: His Olympic education odyssey

Deanna Binder

In 1978, a young German football player from Münster received his doctoral degree (Dr. phil.) in pedagogy, sociology and sport medicine from the Westfälische Wilhelms-Universität Münster. His passion for youth sport and sport pedagogy - as evidenced by 33 years of commitment to these fields as a professor at the University of Essen/Duisberg-Essen from 1980–2013, and his participation as a pioneer in organizations such as the Commission of Sport Pedagogy in the German Society of Educational Science and the International Committee of Sport Pedagogy - meant that when Olympism became a topic of discussion not only for sport historians, philosophers and sport enthusiasts, but also for educators, Professor Dr. Roland Naul became also a pioneer in the emerging field of Olympic education.

> *Roland came across the Olympic Movement, and especially Willibald Gebhardt, in the early 1980s when he was investigating the social history of school-sports/Turnen in imperial Germany. Here he was interested above all in the great German school-festivities, that included gymnastic exercises as a standard programme. The most famous of those school-festivities were the Sedan-Spiele, commemorating die German victory over France in 1871 near the town of Sedan.*
>
> Geßmann, 2018

Of course, this was the same victory that motivated a young Baron Pierre de Coubertin to lament the lack of physical fitness in his French countrymen and campaign for the inclusion of regular games and sport in the schools of France and, ultimately to propose, in 1894, the organization of an international Olympic Games.

This chapter will follow Roland's odyssey through the somewhat fraught waters of the German Olympic Movement as he wove his historical interest in Gebhardt's educational legacy with an evolving commitment to place sport pedagogy front and centre during discussions, conferences, and symposiums exploring Olympism, Olympic values and Olympic educational initiatives. This was not an easy task since the field in the 1980s and 1990s was dominated by (a) those who revered de Coubertin and his ideas and (b) by sport historians, philosophers, sociologists, psychologists and various other academically focused sport scientists. What went on in classrooms or on the sport field with young people and their teachers or coaches was often an unknown and sometimes uninteresting lifeworld for the theoretically focused. Roland's ultimate achievement was to create bridges between these worlds.

5.1 The Willibald Gebhardt project

Naul's first publications appear in the late 1970s and focus through the 1980s mainly on topics related either to curriculum in school sports and physical education or to the history of the evolution of sport education in German schools. In the year of his graduation with a doctoral degree, Roland became a founding member of the International Society for Comparative Physical Education and Sport (ISCPES); in 1989 his first article on an Olympic theme titled 'Olympic headquarters: A new challenge to sport pedagogy in the Federal Republic of Germany' appeared in volume 6 of the ISCPES book series *Comparative Physical Education and Sport* (Fu, F. et al., eds., 1989). This article was a response to the increasing interest by the West German Olympic Committee (NOK) in producing educational materials to foster interest in schools in upcoming Olympic Games and included a challenge to the NOK to highlight sport pedagogy in the learning resources that were being made available to schools. As Roland himself points out, the term *Olympic education* goes back only to the year 1975 when Norbert Müller, a well-known German Olympic historian, published, in German a first article with the topic 'Olympic education' (Naul, 2017, p. 179). Through the 1980s, Naul, who was aware of the educational ideas of Pierre de Coubertin was comparing them with the educational ideas of a German scientist in whom he had developed an ongoing historical interest – Dr. Willibald Gebhardt. In his seminal work *Olympic Education* (Naul, 2008), Naul describes Gebhardt as one of the 'pioneers' of the Olympic Movement. He was the German representative on de Coubertin's original International Olympic Committee (IOC), and after the Athens

Games of 1896 founded the first permanent national Olympic committee. Gebhardt had a profound interest in education; his ideas mirrored many of those of Pierre de Coubertin. According to Naul, in Gebhardt's address of 1904 titled 'The Olympic Movement and the School',*Gebhardt criticised the fact that most academically educated people had previously not recognised 'the educational moral benefits of sports, games and gymnastics'. He argued that 'care of the body is a vital prerequisite for building character in order to develop courage, self-confidence moderation, control of the emotions and other manly virtues'. In this context Gebhardt cited three central functions of an Olympic Education, although he did not himself use the term at that time:*

- the 'social significance of physical exercise', which has a balancing and pacifying effect on the individual
- the national values of physical culture to achieve fitness for military service
- the international function to bring together the various nations by dismantling national prejudices between them and preventing chauvinism (Gebhardt, 1904, pp. 109–110 quoted in Naul, 2008, p. 44).

Naul goes on to suggest that Gebhardt's call for the reform of school education in the context of the Olympic Movement 'reflects criteria and value systems which, at that time, were not generally represented among other members of the IOC' (Naul, op.cit., p. 45). While concurring with his colleagues' pedagogical value system he extended sporting character building to the Olympic sense to include the educational idea of a healthy and active lifestyle. It is only now [as Olympic education initiatives focus on holistic, integrative active living objectives] that we can truly assess and honour this 100-year-old vision (Naul, op.cit., p. 45).

With Gebhardt's inspiring vision in mind, Naul mobilized to establish an institute that would plan and implement research projects to evaluate and promote quality teaching and active sport participation in schools, clubs and communities. One pillar of this work has been a focus on Olympic education. **The Willibald Gebhardt Institut e.V. (WGI) – Forschungsinstitut für Sport und Gesellschaft** was founded on 5 May 1992; it has been pursuing Gebhardt's vision now for more than 25 years. A central mission of the institute is the promotion and protection of the human and social foundations of sport. Although very few people had even heard of Dr. Willibald Gebhardt or his contribution to the German Olympic Movement, certainly very few 'Ausländers', Roland writes:

> *Bei dieser besonderen Zielsetzung für unsere Forschungsarbeiten, war es naheliegend, hier den „Olympischen Gedanken" in den Vordergrund für die Institutsarbeit zu rücken, wie er in Form der „Olympische Prinzipien" in der Charta des Internationalen Olympischen Komitees (IOC) Eingang gefunden hat. Insofern war es naheliegend, das Institut zu Ehren des Begründers der Olympischen Bewegung in Deutschland, Dr. Willibald Gebhardt nach seinem Namen zu benennen, in Erinnerung an seine besonderen Verdienste für die Förderung der Olympischen Idee.*[20]
>
> Naul, 2012b

Today, more than 25 years later, approximately 20 sport and scientific institutions (regional, national and international) belong to the WGI along with a number of researchers from Europe, Asia and North America, some of whom are fellows of the institute. Originally located at the University of Duisberg-Essen, the WGI has recently been reorganized and relocated to Roland's new hub of operations at the Westfälische Wilhelms-Universitaet Münster (WWU). Its new president is Prof. Dr. Maike Tietjens, WWU Münster. With maturity, the WGI is now able to define its work in terms of four competence areas: elite sport, children and youth sport, health and wellness through sport and (unsurprisingly) Olympic education. Each is led by a vice president. Also, unsurprisingly, Roland is the new vice president of Olympic education for the WGI.

20 *This goal for our research work provided a relevant connection with the 'Olympic ideas', which have their form in the 'Fundamental Principles' of the charter of the International Olympic Committee, and to place these ideas in the foreground for the work of the Institute. Therefore it seemed appropriate to honour the founder of the Olympic Movement in Germany by giving the institute his name in memory of his special contributions to the furthering of Olympic ideas (Naul, 2012b, translation by Binder).*

5.2 Olympic education comes of age: The 1990s

With the founding of the WGI in 1992, Roland was well positioned to participate in the escalation of interest and activity in Olympic education in the national and international lifeworlds of the Olympic Movement in the 1990s. A number of factors contributed to this growing enthusiasm.

The 1980s was a decade of curriculum reform in the fields of school sport and physical education in many countries, including Germany. These reforms were highlighted by a shift in emphasis from teaching sport disciplines to using participation in sport/physical education as a mileu for furthering the personal and social development of youth, thus aligning them with the broad educational ideals of Olympism as stated by de Coubertin in his original *Olympic Charter* and with the 'fundamental principles' of subsequent versions of the *charter.* Naul's curriculum/school-based work in the 1980s contributed to this shift, although the debates among sport educators in Germany – East and West - raged until German reunification in 1990. Naul describes this situation as 'the battle of the two German systems' (Naul, 2008, p. 48). Already in 1987, Naul identifies his pedagogical position with respect to this 'battle' with a paper titled 'Sporterziehung als Bestandteil einer neuen Allgemeinbildung'[21] *(Heid & Herrlitz, 1987, pp. 161–171).* This orientation to an integrated approach for sport education also characterizes his evolving ideas on Olympic pedagogy.

The foundation of the International Olympic Academy (IOA) near the ruins of Ancient Olympia, Greece, in 1961 also contributed significantly to a growing understanding of the central role of education and pedagogy in the promotion of Olympism and the Olympic Movement. Before then, as Müller in 1975 succinctly noted, 'In spite of its worldwide significance, the Olympic Movement to which Coubertin assigns [a significant] educational task, is no more able to fulfill it than was Coubertin half a century ago' (Müller, 1975, p. 138). The IOA, supported by the NOC (National Olympic Committee)

21 *Sport education as a component for a new integrated approach to education.*

of Greece, offered a peaceful academic setting for lectures, discussions and activities among participants from all the nations with membership in the IOC. Müller was among the very first participants at the IOA, at a time when the facilities were hosted under canvas.

> *Ancient Olympia, located in the remote and lovely valley of the river Alpheios on the west coast of the Peloponnesus is a paradise, a site of great religious importance since prehistoric times. Two thousand years ago, it took two months of travel for athletes and their retinues to get from anywhere else to this spot. Greek slaves considered a trip to Olympia as punishment. Carrying living materials and provisions enough to camp out for a month on the hot plains of the river valley was no one's idea of a good time. Today, a five-hour bus ride from Athens brings delegates to the Academy, which is pleasantly situated just around the corner from the excavations. Echoes of voices of ancient times drift through the olive and pine trees, and are especially present as you pass the moonlit excavations of the ancient stadium and the nearby sacred precincts on a thirty minute walk to a taverna in the village.* Binder, 2002, p. 74

Gradually, over a period of decades, participants who were influenced and informed by their experiences at the IOA disseminated a new awareness about Olympic education to their national Olympic networks. Naul has also frequently experienced the magic of the IOA – as a participant in various sessions, and as a professor/lecturer, most recently as a guest professor teaching the Olympic pedagogy module for the master's degree programme in Olympic studies offered by a partnership between the University of the Peloponnesus and the IOA.

Olympic Games host cities have stimulated interest in Olympic education through projects for schools. Tokyo (1964), Munich (1972), Montreal (1976) and Los Angeles (1984), for example, had regionally based initiatives to provide information about the Olympic Games to their citizens, including students in schools. However, as Naul et al. (2017, p. 20) point out,

'it was the nationwide success of the Calgary 1988 Olympic Winter Games school kits, titled Come Together: the Olympics and You (Binder, 1987) which attracted the attention

of the IOC and focused its awareness on the potential of integrating Olympic themes, values and ideas with regular school curricula.'

Subsequently, at its 1994 Centennial Congress, the IOC required that cities bidding for an Olympic Games include a section describing plans for an education programme in their bid submissions. Based on the Calgary 1988 Winter Games model, Albertville (1992), Lillehammer (1994), Sydney (2000), and most subsequent Games organizing committees have provided comprehensive curriculum-based educational materials and programmes as part of their Olympic communication strategies. Thus, although it was still academically under-theorized, Olympic pedagogy was now on the agenda for the field of Olympic studies. And although Norbert Müller (2000), a renowned Coubertin scholar and Olympic historian, was still leading the Olympic education endeavours of the German Olympic Movement and writing more and more about the pedagogy of Olympic education, educationalists such as Naul (2002, 2003, 2006), Geßmann (1992, 2002, 2004) and Grupe (1993, 1996, 1997) were doing the groundwork in Germany for the creation of a theoretical framework for Olympic pedagogy.

Roland's contribution to this work, carried out through the research projects of the WGI during the 1990s, involved, among other things, a series of baseline surveys to compare the physical abilities and lifestyles of young people in Germany and then in five other European countries. Those studies were based mostly upon questionnaires, and they often included an 'Olympic questionnaire'. Thus Roland and his international colleagues wanted to analyze (a) what young people know about the Olympics and the Olympic idea and (b) their attitudes and values related to the values of the Olympic Movement, such as fair play. The results were published in many journals, congress reports (above all in International Council for Sport Science and Physical Education [ICSSPE] studies and reports) and in chapters in some books up until about 2002.

A detailed summary of the results of these studies is beyond the scope of this chapter. However, using Roland's words the following are some highlights from this work of the '90s.

> *The outcomes of these studies show in summary a poor standard of knowledge about Olympic-related information. The findings suggest that the older the stu-*

> *dents are the more they know about it; boys know more than girls. Pupils know more about recent Olympic events . . . this is not surprising because the majority of school students take their knowledge from TV and other electronic media . . . only a very few reported that they received Olympic information in school or from their sport coaches.* Naul et al., 2017, p. 185

With respect to fair play, Naul (2003) notes that *fair play* as a concept is strongly supported by 12- to 15-year-old German youth. However, 25 per cent of them are not sure that *fair play* is a value practised during the Olympic Games, and a third of them suggested that they would respond unfairly in practice on the sport field if confronted by an opponent. Naul suggests that this dissonance between idealism and practical reality offers a good reason to introduce Olympic values education within the context of school sport (Naul et al., 2006, p. 252).

During these years, Roland's commitment to collaboration is evidenced by the establishment of a long-time affiliation with Professor Antonin Rychtecky of Charles University in Prague, Czech Republic, with whom he carried out a number of comparative studies, published research (Naul & Rychtecky, 2000), and with whom to this day he has continued a close working relationship. In 2017, Charles University in Prague awarded him an honorary doctorate in recognition of his decades of support for its research programme in physical education and Olympic studies.

5.3 Intersections

Roland's Olympic education odyssey and mine intersected often in the years following the turn of the century. By way of introduction, briefly, my Olympic baptism occurred in Calgary in 1984 when I became the supervisor, education for the Calgary 1988 Olympic Winter Games. *Come Together: The Olympics and You* was my project, and its success became my ticket for a change in career from teacher and curriculum developer to Olympic educational consultant, for numerous appearances at the IOA, and for the challenge and privilege of developing two international Olympic education manuals, *Be a Champion in Life,* produced by the Foundation of Olympic and Sport Education (FOSE) in Athens, Greece, 2000, and *Teaching Values: An Olympic Education Toolkit* – the Olympic Values Education Programme (OVEP) of the IOC, 2007. As an Olympic educator with a speciality in curriculum development and pedagogy sooner or later I was going to meet Professor Dr. Roland Naul.

Info

Athens, Greece, October 2001

Dr. Margaret Talbot and I were hosting a workshop on Be a Champion in Life (2000) prior to the IAPESGW (International Association of Physical Education and Sport for Girls and Women) Congress of 2001 taking place in Alexandria, Egypt. I had written Be a Champion in Life: A Teacher's Manual with the collaboration of the Athens FOSE between the years 1997 and 2000. The president of FOSE, a wealthy Greek of the old school, launched it in 2000 at a conference at the foot of Mt. Olympus. Its integrated pedagogy was formatted around five educational values of the Olympic Movement that are highlighted in the Fundamental Principles of the Olympic Charter (joy of effort in sport and physical activity, fair play, pursuit of excellence, respect for others and balance in life). Drafts of the manual had been piloted globally in classrooms on five continents. Margaret Talbot, who was then the president of IAPESGW and a vice president of ICSSPE was on my advisory group. Since our workshop was taking place exactly a month following the horrific events of 11 September 2001, in New York City, half the world was in lockdown; many congress participants either couldn't or decided not to attend the conference in Egypt. We didn't expect much of an attendance at our workshop in Athens, and since IAPESGW was a 'women in sport' conference, we were surprised by the appearance of a man at our workshop. I have never asked Roland how he found out about the workshop or why he decided to attend, but the outcome of his attendance and of our subsequent discussions was a shared conviction that Olympic education as values-based education resonates in varieties of learning environments and that a key pillar of a pedagogy of Olympic education is not necessarily 'what to teach' but 'how'.

The first iteration of Roland's thinking on the topic of the 'how' – what I call learning and teaching methodology and what he calls 'didactics' – appears in the section on education in the *White Book on Olympic Education* that accompanied the Düsseldorf Rhein-Ruhr 2012 Olympic Games bid submission (Naul, 2002, 2012a). 'The three parts of the [White] book contained a new, **integrative** concept of Olympic education, a list of a variety of Olympic education events scheduled for kindergarten schools, sport clubs, sport association higher learning institutes and public authorities including seminars for sport journalists. The *White Book* also included a bibliography on scientific and practical materials to study and to teach Olympic education' (Naul et al., 2017, p. 184). His emerging thoughts on didactics are also explored in another 2004 publication titled *Von der Pädagogik des Olympismus zur Didaktik der olympischen Erziehung* (From the pedagogy of Olympism to the didactics of Olympic education) and are then fully articulated in his most important work, *Olympische Erziehung,* 2007 (*Olympic Education,* 2008), a book which is in high demand as the current go-to text for Olympic educators and which has now been translated into Japanese.

Info

Düsseldorf, 2002

I have just graduated with a PhD in Curriculum Studies from the University of Alberta and jokingly referred to it as my 'retirement project.' After ten years as an entrepreneur in a curriculum development consultancy I have established, with the support of the dean of the Faculty of Education, an International Institute for Olympic Education at the university with a signed collaboration with the Canadian Olympic Committee. My vision is to create a global communication network for Olympic educators and to focus on values education and Olympic pedagogy. Like Roland, I have had to swim upstream to gain recognition for the field of curriculum and pedagogy in the world of Olympic studies. At the university level the academics in the humanities sometimes show little respect for the professionals in applied fields such as education. I want this to change. So does Roland. He has already spent 20 years swimming upstream in the same stream. In the spring of 2002, I receive a letter from the Düsseldorf Rhein-Ruhr 2012 GmbH, the committee organizing the bid for the city for the 2012 Olympic Games. A Mr. Hartmut Gustmann, an official of the bid committee, is inviting me to make a presentation on my Olympic education work – specifically the work on *Be a Champion in Life* – to a public forum being held to support the bid process. Of course, Roland is behind this invitation, as he is behind so many other invitations that I have received since then to attend Olympic education-related events sponsored or supported by the WGI in Europe. I attend the forum and make my presentation – my first-ever in the German language. I can speak and read it, but to present or to write in German is a terrifying undertaking. It goes well, they say. I spend a delightful week in Düsseldorf with German colleagues, whom I have met over the years at the IOA and at conferences. I am entranced by the city. Its motto is 'So bunt wie die Welt', and the city is covered in colours with posters everywhere of faces of every culture. From my point of view, they can't lose. The forum also provides opportunities for discussions about a tentative collaboration with Roland. We talk about linking the work of our two Institutes and maybe creating a multi-institutional master's degree in Olympic Studies with a specialty in Olympic pedagogy. What an opportunity!

My presentation goes well, but the Düsseldorf Rhein-Ruhr Olympic bid does not. Leipzig is selected as the German city to bid for 2012, and it drops out too; London gets the Games. But that is not the end of the story of the *White Book*.

> *The draft of the White Book became a platform for the development of a new Olympic education concept for Germany with two essential components: first the traditional historic-philosophical dimension was supplemented by a modern orientation to the Olympic Movement, and second a more comprehensive and integrated view on different didactic demands for teaching Olympic education was articulated.* Naul et al., 2017, p. 184

Experiential Moments

Edmonton, July 2004 – The International Institute for Olympic Education at the University of Alberta is holding its third Summer Institute. Twenty students from a variety of disciplines will participate in a five-day experiential exploration of the Olympic Movement and Olympic values. The guest professor is Roland Naul. He experiences the floods of a spring thaw in Edmonton. I become a fellow of the WGI.

Thessaloniki, 2004 – Roland and I present our first combined paper at the ICSSPE 2004 Pre-Olympic Congress: 'Pedagogy and Olympic Education: A Trans-national comparison between North America and Europe'.

Sportschloss Velen, November 2007 – At a WGI-sponsored symposium on Olympische Werte Erziehung in Schule und Verein I make my first presentation on the IOC Olympic Values Education Programme – The Olympic Values Education Project of the IOC (OVEP). The setting is idyllic. I feel like a princess in my room in the castle; a black swan swims in the moat under my window. Roland has just published his book *Olympische Erziehung.* He agrees to act as a primary reviewer for drafts of *Teaching Values: An Olympic Education Manual* (the OVEP handbook).

Frankfurt, October 2012 – A Presentation for the Conference of the German Olympic Academy for Europe's National Olympic Academies. It is an honour for a North American to be there. I talk about the global initiatives to implement OVEP. By now I am travelling regularly to conduct workshops, for example, in Singapore, Trinidad and Tobago, India, Tanzania. Roland is also travelling, bringing pedagogy to the field of Olympic Studies. The IOC authorizes work on a revised edition of the OVEP programme.

Istanbul, December 2013 – ICHPER (International Council for Health, Physical Education, Recreation, Sport, and Dance) Congress. I suggest to the organizers that any symposium on Olympic education feature Roland on the podium. He attends along with colleagues from Zambia and Colombia. I outline the insights over three decades that ground my orientation to Olympic education in a presentation titled 'Evolution of a Pedagogy for Olympic Values Education'. We begin the planning for a comparative international review of the Olympic education activities in different countries of the world.

February 2017 – We publish our book, *Olympic Education: An International Review*, with Routledge after many years of planning, translating, analyzing, writing, editing and re-editing. The book includes chapters from 20 countries, and is launched at a special WGI Conference in Prague, Czech Republic, in May 2017 with some of our authors in attendance. Roland's long-time colleague Antonin Rychtecky is therefore able to attend although his health is failing. Roland honour's his work.

5.4 Olympic education: the book

The above excerpts offer examples of the ways in which Roland's Olympic education work and in particular his focus on pedagogy and mine have intersected. Hopefully, they will illuminate some of the experiential moments. My work culminated in a teaching manual/curriculum resource (Binder, 2007) that helps to guide the development of Olympic values education for the IOC. Roland's contributions to that project were invaluable. Roland's work culminated in the book *Olympische Erziehung* (2007) and in its English equivalent, *Olympic Education* (2008). It is this book that has cemented his reputation as one of the world's premiere Olympic education theorists. It comprehensively bridges the gap between theory and practice. In 5 parts and 15 chapters, the book shows why the Olympic ideals are a modern challenge not only for a new physical and sport education but also for the development of essential life skills for today. It introduces pedagogical and didactical fundamentals for an Olympic education, in order to bring motor abilities, social behaviour and moral actions in sports and everyday life back together again – not only in the mind, learning and actions of children and youth, but also that of grown-ups in the social settings where young people live.

After dealing with the history and philosophy in that book, he lays out the various concepts of Olympic education that crowd each other in the literature and then outlines a simple, useful and now well-used analysis of the approaches to teaching Olympic education. Chapter 11 features a description of the various pedagogical concepts, assigning them to their various proponents. For example, Olympic education in the spirit of Pierre de Coubertin is the orientation of Norbert Müller; Olympic education as a cultural task for school sports is a concept discussed by Omno Grupe; Olympic education as a didactic objective for physical education is Rolf Geßmann's orientation and Olympic education as value education for everyday life, Naul suggests is Deanna Binder's orientation (pp. 102–112). Some of the discussion in this chapter is a bit fuzzy, perhaps because there is too much material to try and summarize. However, the analysis that he draws from this summary for chapter 12: 'Didactic approaches to teaching Olympic education' is crystal clear. He describes four teaching approaches and thereby defines the field:

- the knowledge-oriented approach
- the experience-oriented approach
- the physical achievement – oriented approach
- the lifeworld-oriented approach

Olympic educators the world over now use this analysis in their discussions of topics related to Olympic education. He also describes a fifth approach which he describes as 'the integrated approach' (pp. 122–130). His thinking on this topic is still a work in progress, but useful nonetheless as a point of departure for future discussions about Olympic pedagogy.

Since the publication of this book, Naul has been sought after globally as a speaker and consultant for Olympic education-related endeavours. He participated with the National Olympic Academy of Brazil in the development of their materials for the Rio 2016 Olympic Games. He is in ongoing communication with the officials of the organizing committee in Tokyo who are responsible for Olympic education initiatives. *Olympic Education* was recently translated into Japanese.

5.5 Not a Conclusion

For Professor Dr. Roland Naul there will never be a 'conclusion'. He is a tireless worker and effective administrator and relentless in his passion for Olympic education and youth sport. The WGI has a number of ongoing research projects ongoing, most of them now relating to the obesity crisis and focusing on challenges of promoting 'healthy children in sound communities'. Within its new administrative form, Roland, as vice president for Olympic education for the WGI will doubtless be engaging in new initiatives to assure that Olympic education becomes an integrated component of school sport and in the sport clubs. His curriculum vitae documents affiliations with dozens of organizations past and present, and no doubt he will be carrying on with some of these.

This chapter has attempted to bring to life his odyssey in pursuit of a pedagogy for Olympic education. What it hasn't documented is his odyssey within the German Olympic Movement itself, which, over the years, has undergone a number of transformations. That story can be told by others closer to the scene. My understanding is that it is a story fraught with personal and administrative challenges, as is always the case when convictions confront bureaucracy. I hope that I have done justice to his journey and thank him profoundly for all his contributions to mine. Maybe we haven't quite come home to Ithaca, but we are getting close. Best wishes.

References

Binder, D. (1986). *Come together: The Olympics and you (Elementary, Junior High School and Senior High School Resource Manuals).* Calgary: Organizing Committee for the 1988 Olympic Winter Games.

Binder, D. (2000). *Be a champion in life: An international teacher's resource manual.* Athens: Foundation for Olympic and Sport Education.

Binder, D. (2001). "Olympism" revisited as context for global education: Implications for physical education. *Quest, 53*, 14–34.

Binder, D. (2002). *Curriculum odyssey: Facilitating an international Olympic education project.* Unpublished PhD dissertation, University of Alberta.

Binder, D. (2007). *Teaching values: An Olympic education toolkit.* Lausanne: International Olympic Committee.

Geßmann, R. (1992). Olympische Erziehung und ihre schulische Umsetzung. In NOK für Deutschland (Ed.), *Olympische Erziehung in der Schule unter besonderer Berücksichtigung des Fair play-Gedanken.* Frankfurt: NOK , pp. 33-43.

Geßmann, R. (2002). Olympische Erziehung in der Schule. Zentrales und Peripheres, *Sportunterricht, 51*(1), 16–20.

Geßmann, R. (2004). Olympisches Menschenbild und schulische Sportdidaktik. In NOK (Ed.), Olympische Erziehung. *Eine Herausforderung an Sportpädagogik und Schulsport* (pp. 131–153). St. Augustin: Academia.

Geßmann, R. (2010/11). Olympic education: Fair play and their practice in schools. In Play Fair! Academic Supplement. Issue No. 8, *The Official Publication of the European Fair Play Movement 2010/2–2011/1*, 2–4.

Geßmann, R. (2018). *Email correspondence with D. Binder.* March 3, 2018.

Grupe, O. (1993). Olympisches Menschenbild und olympische Erziehung. In R. Pohl (Ed.), *Facetten der Sportpädagogik. Beiträge zur pädagogischen Diskussion des Sports.* Schorndorf: Hofman, pp.31-38.

Grupe, O. (1996). Die Olympische Idee ist pädagogisch: Zu Fragen und Problemen einer ‚olympischen Erziehung. In N. Müller & M. Messing (Eds.), *Auf der Suche nach der Olympischen Idee.* Frankfurt: Agon-Sportverlag, 23-38.

Grupe, O. (1997). Olympismus und olympische Erziehung. Abschied von einer großen Idee? In O. Grupe (Ed.), *Olympischer Sport – Rückblick und Perspektiven.* Schorndorf: Hofmann, 223-243.

Heid, H., & Herrlitz, H. (Eds.). (1987). Allgemeinbildung: Beiträge zum 10. Kongreß der Deutschen Gesellschaft für Erziehungswissenschaft. *Beiheft der Zeitschrift für Pädagogik No. 21.* Weinheim.

Naul, R. (1987). "Sporterziehung als Bestandteil einer neuen Allgemeinbildung". In H. Heid & H. Herrlitz (Eds.), *Allgemeinbildung: Beiträge zum 10. Kongreß der Deutschen Gesellschaft für Erziehungswissenschaft. Z. F. Päd.*, Beiheft 21. Weinheim, pp. 161-171.

Naul, R. (1989). Olympic headquarters: A new challenge to sport pedagogy in the Federal Republic of Germany. In F. Fu, et al. (Eds.), *Comparative physical education and sport, Volume 6 of the ISCPES book series.*

Naul, R. (2000). Olympische Ideale: Idee und Wirklichkeit aus der Sicht von europäischen Jugendlichen. In N. Müller & M. Messing (Eds.), *Blickpunkt Olympia. Entdeckungen, Erkenntnisse*, Impulse. Kassel: Agon, 175-190.

Naul, R. (2002). Olympische Erziehung – Aufgaben und Chancen für den Schulsport. *In Sportunterricht, 53*, 267–272.

Naul, R. (2003). *Fair Play as an Olympic ideal and a personal behaviour pattern in sports activities: Vision and reality assessed by young European students.* Online paper. Retrieved March 6, 2018, from http://www.olympischeerziehung.de/media/Fair_Play_as_an_Olympic_Ideal_and_a_Personal_Behaviour_Pattern_in_Sports_Activities_Vision_and_Reality_assessed_by_young_European_Students.pdf

Naul, R. (2004a). Die Dekade der Olympischen Erziehung an Rhein und Ruhr. In Düsseldorf Rhein-Ruhr GmbH & Deutsche Sporthochschule Köln (Ed.), *Olympic Spirit*. Köln: GfP, 72-86.

Naul, R. (2004b). Von der Pädagogik des Olympismus zur Didaktik der olympischen Erziehung. In NOK (Ed.), *Olympische Erziehung – eine Herausforderung an Sportpädagogik und Schulsport*. St. Augustin: Academia, 113-130.

Naul, R. (2007). *Olympische Erziehung*. Oxford: Meyer & Meyer.

Naul, R. (2008). *Olympic education*. Oxford: Meyer & Meyer.

Naul, R. (2012a). Olympic pedagogy – A new discipline of research and teaching at Olympic Study Centres? University of Loughborough, Centre for Olympic Studies and Research (COS&R) (Ed.), *Proceedings of the First International Colloquium of Olympic Study and Research Centres*, July 2012, 11–12.

Naul, R. (2012b). *Das Willibald-Gebhardt-Institut e.V. – ein internationales Forschungsinstitut für Olympische Erziehung. A presentation to the Global Forum for Physical Education Pedagogy.* Velen, Germany. Retrieved March 2, 2018, from http://www.globalpeforum2012.org/uploads/media/wortmann_250775.pdf

Naul, R. (2012c). Olympische Erziehung: Ein integriertes Konzept für Schule und Verein. In Düsseldorf Rhein-Ruhr 2012 GmbH (Ed.), *Weißbuch Olympische Erziehung*, B1–B32. Düsseldorf.

Naul, R., & Rychtecky, A. (2000) Kenntnisse, Schulunterricht und Informationsquellen zum Thema Olympia bei deutschen und tschechischen Jugendlichen. In Deutschen Olympische Institut (Ed.), *Jahrbuch des Deutschen Olympischen Instituts, 1999.* St. Augustin: Academia.

Naul, R., et al. (2006). *Olympische Werteerziehung in Schule und Sportverein. Handreichung mit Materialien für Multiplikatoren, Moderatoren und Bildungsreferenten für den Schul- und Vereinssport.* Essen: Willibald-Gebhardt-Institut.

Naul, R., Geßmann, R., & Wick, U. (2008) *Olympische Erziehung in Schule und Verein.* Grundlagen und Materialen. Schorndorf: Hofmann.

Naul, R., et al. (2017). *Olympic education: An international review.* London: Routledge.

Telama, R., Naul, R., Nupponen, H., Rychtecky, A., & Vuolle, P. (2002). *Physical fitness, sporting lifestyles and Olympic ideals: Cross-cultural studies on youth sport in Europe.* Schorndorf: Hofmann.

6 More than a minor matter: Roland Naul and his merits in the field of soccer

David Niehues

With regard to all the work Roland has done throughout the course of his academic career, his activities in the field of soccer, both on a theoretical and on a practical level, should not be compared to his accomplishments in the field of physical education in youth or Olympic education. But everyone who knows Roland is aware of the fact that soccer as a sport itself as well as the concepts within and learned by playing soccer have always had a high meaning for him. Thus, his work in the field of soccer is worth more than just a note and will be the topic of the following pages.

When asked about his sporting 'roots' as a youngster and student, Roland will always refer to the sport of soccer as something that shaped both his physical skills and, more importantly, his way of thinking and working. The sport of soccer in all its facets is a reoccurring element in the life of Roland Naul: as an adolescent playing soccer, as a professor in both teaching and research, and also in his personality, sharing values and beliefs he learned from the sport over the years like comradeship, persistence and serenity.

Maybe his passion for the most famous sport in Germany has its roots in the place and time he grew up and was educated in, both socially and from a sporting perspective. Born in 1948, Roland was at the age of an elementary school kid, when he witnessed one of the most outstanding and until today extraordinary achievements of the young German republic in the field of sport: The 'Wunder von Bern', with the German soccer national team winning the world cup against an apparently invincible team from Hungary. It is worth a note that Helmut Rahn, arguably the most remembered hero from the world cup win and the striker scoring the winning goal in the final, played for the soccer club RW Essen in the Ruhr area at this time. As it turned out later, Roland would work and partly live for over 30 years at the University of Essen, in which one of the nation's idols played soccer during his upbringing. Roland grew up in Bocholt, a city not far from the Ruhr area, which was famous for the most prominent soccer rivalry in Germany. Retrospectively, considering his time as a soccer player, mainly playing for his hometown club FC Bocholt, and his work in the academic field of soccer as a professor, it can be stated that his merits on the pitch could not live up to those off the pitch. Despite never being his main area of work, not in research and not in teaching, Roland tried to pass back as much to the sport of soccer as it gave him while growing up.

With over 30 publications from 1986 to 2010, his commitment to the sport of soccer never ended up and can surely be entitled as persistent. Instead of a chronological accumulation of work he has done in and around soccer, this chapter will subdivide his work into four major sections that can be identified when investigating Roland's work in the broad field of soccer. Each field will be introduced using exemplifying work Roland has done over the years:

(1) soccer in school, (2) soccer in academics, (3) soccer in sport clubs and associations, (4) women's soccer. The following sections will tell the story of Roland's – in every way – greatest pastime in the world: soccer.

6.1 Soccer in school

Roland has always been an advocate of freedom of play when it comes to teaching soccer in school. In one of his first works regarding the history and development of soccer as a school game, he describes the transition of soccer from a game primarily played by students in a variety of social settings (in break times, after school, etc.) to a game which forces students to practice specific details in inappropriate and not play-oriented settings. From a historical point of view, soccer only found its way into the curriculum through control and ritualization by physical education teachers. Or in a different manner, the more rules the game of soccer entailed, the more appropriate it became for being taught in school. In fact, Roland claimed that the way soccer was taught in schools is detrimental to what the game entails and needs in essence players who are creative by their own personal style of problem-solving competence (Naul, 1986). The phenomenon Roland described in 1986 can be adapted to the current debate about the athletic education in professional youth academies (German: NLZ). The result of the German national team at the World Cup 2018 led to a public debate about the conformity throughout youth soccer education and the lack of creative players with unique characteristics in Germany. Recent studies have shown that creativity in soccer leads to better team performance (Kempe & Memmert, 2018). Maybe the conclusion Roland drew 30 years ago just needed the public attention it gets currently. Roland's suggestions are in line with research that suggests open-play as a better tool to develop sport-related creativity than the simple repetition of a certain movement or skill (Memmert, Baker & Bertsch, 2010). Roland identified this issue, be it in a slightly different educational form, as one of the firsts years ago.

It is not a surprise that most of his work in school soccer is related to the different developmental stages soccer has passed through over the years. As someone growing up in the young German republic, he had first-hand experience with soccer being taught in school. In one of his later works he outlines the ambivalent development of the physical education curriculum regarding soccer: At first, soccer was suggested being a tool for character building, facilitating the development of a variety of personality traits such as companionship, responsibility, agreeableness or fairness. Several reforms made soccer a quasi-mandatory sport that was taught on an advanced level with regard to the content that was taught: among others, individual and team tactics, role behaviour and different playing styles were on the curriculum for boys aged 15 or older. Several years later, soccer was relegated from its exceptional position as one of the main team and ball sports in the federal state. In fact, soccer is only one of many sports games in the modern curriculum of elementary and secondary schools. The sport-oriented curriculum was replaced by a movement-oriented one, where the particular sports were no longer attractive. In this case, the skill of 'manipulating the ball with the foot' rather than playing soccer as a sport was the main idea of the sport educators (Naul, 2006; Naul, 2010). In a never published transcription of the 'DFB-Masterplan: Team Schulfußball 2014' that was drafted after the enormous success of the world cup 2006 in Germany, Roland suggested a three-level plan to further promote soccer in schools. Besides structural changes in terms of renewing facilities and equipment in schools as both short- and long-term actions, he also encouraged a closer connection to schools and educational universities to improve the technical education of teachers within soccer. Ultimately, there should be coordinators for school soccer to ensure a better exchange between the basis, the schools in which soccer is played, and sport clubs or associations.

6.2 Soccer in academics

A rather small, but not marginal, note in Roland's work aimed at the soccer education at universities in Germany: As we saw in the previous subsection, Roland was highly engaged in the development of the physical education curriculum when it comes to teaching the game of soccer in school. Not less than his efforts in teaching pupils in school were his efforts in enhancing the education of students, especially those who studied with the aim of becoming a physical education teacher in the future. Back then, soccer was a mandatory course in all physical education teacher studies, with different amounts of hours taught per week depending on the university. In his evaluation of a questionnaire distributed at a symposium for physical education teacher education in 1985, he defines similarities and differences in regional education for students. He outlines regional and structural differences, aiming at the establishment of a shared main curriculum for teaching sports at universities. A recurring contrast between different universities refers to the aim of teaching soccer at universities. What is the focal orientation point? Is it the education of future teachers or should it rather be the properties of the sport soccer itself? How should the incorporation of women into the educational system for teaching soccer look like (please note that in 1985, co-education was still open to debate and not an established system at universities in sport)? To what extend should the academic teaching of soccer and the training in local soccer clubs overlap? Or is there a need for greater differentiation? Roland encouraged the discussion about teaching soccer at universities towards a standardization of the curriculum (Naul, Schmagold & Wadle, 1987). In an essay about curriculum development in schools and universities regarding soccer, Roland worked out the alteration of the subject sport under the influence of the Bologna process. Several new disciplines emerged, which predetermined an increased number of hours spent in the subject soccer, whereas in classical physical education teacher education, soccer education was further diminished. To conclude, soccer made its way from a mandatory subject at German universities to a possible option within the scope of general modules entitled sports games (or similar). Concluding and combining his work of sport pedagogy both in school and in the university setting, Roland advises a more individual and differentiated education, especially for future physical education teachers. Not only in soccer, but also in other fields of sports it might be useful to have experts within a certain field in the teaching staff of a school. Thus, a more individualized and sport-oriented physical education lesson could be ensured. For the future, Roland induces the German soccer association (henceforth abbreviated as DFB) to think of the possibility of implementing an own study programme following examples from other European countries, who successfully introduced 'Football Studies' (Naul, 2010).

6.3 Soccer in sport clubs and associations

With the beginning of the twenty-first century, Roland dedicated his work in soccer more towards the organized professional soccer system in Germany. Besides holding several positions within the scientific and professional soccer system in Germany (e.g., member of the school football commission of the DFB), he and his team developed a performance diagnostics tool, that was introduced in 2004 to youth players in North-Rhine Westphalia: The German Soccer Fitness Test (henceforth abbreviated as DFFT for Deutscher Fußball Fitness Test). The aim of this test was to both assess the soccer fitness of youngsters as well as monitor their development over time. The DFFT is composed of altogether six motor test items and a BMI assessment. The test items were partially taken from other test batteries and partially self-elaborated. The 'soccer run' as an item to test speed and agility was elaborated by Roland, incorporating quick changes in direction and 180-degree turns. Other items used in the DFFT were trunk-lift, shuttle run, curl-ups, jump and reach and sit and reach. The fitness components covered with this test were speed, aerobic endurance, strength endurance, power (in sprint and jump) and flexibility. As mentioned before, it was Roland's firm belief that soccer was not only a game of the body and the ball, but rather a game that was also played with the head and ultimately with the brain. Thus, he also implemented a battery of personality assessments together with the DFFT, to test players for (1) self-efficacy in soccer, (2) physical self-concept, (3) ego and task orientation in soccer and (4) fear of failure/self-confidence (Naul, 2005; Naul et al., 2006).

The DFFT and the personality assessment were first conducted as a pilot with a representative team from the Westphalian soccer and track and field association (abbreviated as FLVW) in 2004. Seventy-eight boys who were selected beforehand as promising talents from their local soccer clubs took part in the assessments. After the pilot was conducted, Roland and his team monitored the players for another two years to track their development. Overall, the results indicate that the talented soccer players outperformed their peers in most of the motor components except for flexibility. Although the psychological self-assessment of most talents was in line with their actual motor results, there were some players who highly under- or overestimated their motor performance. This was a clear hint towards specific games that facilitate self-consciousness and a better awareness of personal strengths and weaknesses for the youngsters (Naul, Brüggemann, Völz, 2008a,b; Naul et al., 2008).

An unpublished post hoc analysis of the assessments from 2004–2006 revealed that especially the psycho-mental qualities of the soccer players determinate future career success. Of the 78 soccer players born in 1991 and tested in 2004, 13 made it to the elite level (here: first- and second-tier German or internationally comparable soccer). The only significant difference between the 13 professional players and the 65 remaining soccer players was found in their perceived self-efficacy and fear of failure. Today's professionals showed significantly higher levels of perceived self-efficacy and lower levels of somatic anxiety compared to their peers (Niehues & Naul, 2017). Nowadays it is common practice that every professional soccer club and every representative team of the soccer associations includes performance diagnostics and monitoring in their regular training. There are several test batteries out there and the different clubs or associations rely on different diagnostics, but the importance of close monitoring, especially in youngsters, is a rather new phenomenon compared to the history of the sport itself. After restructuring the talent development programme in 2002, the DFB implemented first elements of performance measurements in its regional bases in 2004 (Lottermann et al., 2003). The work done by Roland and his colleagues was a further contribution in stressing the importance of performance measurements to ensure later excellence in professional soccer.

6.4 Women's soccer

In today's world, it is not possible to imagine that the majority of sport games is reserved for only men or women. The declining prevalence of gender stereotypes and a new gender awareness distinguish the twenty-first century from the times before, also in the field of sport. With that said, it might be astonishing, that soccer, undisputedly the most popular sport in Germany, has a very young tradition as a sport for women. After the world cup win in 1954, the DFB announced a general prohibition for the sport clubs to offer women to play soccer. It took the DFB 15 years until 1970 before soccer was allowed as a women's sport in Germany. However, this was not done without restrictions: cleats were forbidden, balls were smaller and lighter, one match was 70 minutes and women had to pause their season for at least five months during the winter break. With ongoing changes, foremost a loosening of the restrictions, women's soccer gained more and more popularity within society. However, in the field of sport science, women's soccer was highly underrepresented. In fact, one ought to think women's soccer was ignored by a large part of the scientific community. In 1989, the German national team won the first European championship and gained public attention. The same year, Roland Naul, in his position as a professor for Sport Science and Sport Pedagogy, published 'Sportwissenschaftliche Analysen zum Frauenfußball' (Naul, 1989). This work exposed the research done – or better: not done – in the field of women's soccer up to 1989. First analyses conducted in England were characterized by a socio-economic approach towards women's soccer: prevalence and reasons for women's participation in soccer seemed to be the most interesting concerns of research, along with physiological differences such as risk of injury between men and women. A first study that was related to sport science in the narrower sense was done in Australia, where researchers conducted performance tests and demanded better training schedules for Australia's female soccer players. Roland's interest, when entering the field of women's soccer, was different: he approached it in both a socio-economical and cross-cultural way, as well as from another perspective by investigating how women perceived differences between female and male soccer. In the study, he first asked a cohort of female soccer coaches and later a cohort of female students interested in practical soccer education if and how much they associated physiological/motor attributes and in contrast psychosocial attributes to male soccer, female soccer, or both on equal terms. The results indicate that female soccer creates its own profile, which is far from the feminization that was intended by many people back then.

Female soccer was perceived as androgyne in a sense of equal potentials in endurance, speed and athleticism and had its own touch in an increased level of fairness, self-control and elegance. This work, in a field that was merely discussed back then, shows that Roland's interest in making women's soccer socially accepted was based on the assumption that women's soccer has its own characteristics and attributes, especially for the players themselves (Naul, 1989). In a cross-cultural assessment in cooperation with his fellow Soichi Ichimura, Roland expanded this work by comparing different cultural communities for women in soccer. The observed cross-cultural differences between German and Japanese female soccer players are based in a completely different socialization in sport. Sports for young Japanese women traditionally did not include body checking or physical contact at all. Thus, it is no surprise that soccer was perceived more intense in strength and aggression by Japanese women. In contrast, fairness, a characteristic highly important for German soccer players, had no soccer-related meaning at all for Japanese girls, as it is *the* basic value in physical education in Japan (Ichimura & Naul, 1991). These attributions as well as the cross-cultural differences implied that further work was needed to understand underlying processes related to the implementation of women's soccer, especially with regard to cultural differences. More recently, quite some time after the struggles of acceptance in women's soccer, Roland issues chances and problems in the process of talent development for girls in Germany. Combining his experience in the fields of both educational (school) and organized (sport clubs) sport, he elaborates on what the next steps have to be to close the lines between boys' and girls' soccer, both institutionally and on the educational level. In sum, promising soccer-playing girls should undergo the same advantageous procedures as boys do. Cooperation between schools and sport clubs (their NLZs) should also be established for girls, performance diagnostic should be applied and soccer specific-courses in all-day schools should not back off from being offered in a coeducational manner (Naul, 2007).

6.5 Concluding remarks

Recapitulating all the work Roland has done, one particular facet becomes clear: His interest in the sport of soccer was neither limited to the elite nor to soccer in school nor any other specific. Roland always tried to keep a holistic approach regarding conclusions he has drawn about how to change, develop or facilitate certain processes or structures in the sport of soccer. Especially, his comprehensive analysis of the sport of soccer as a leisure-time and performance-oriented sport in Europe is an example of his extensive range of work. He pointed to the exceptional position soccer holds within the leisure-time profile of German kids compared to other countries and unfolds the revival of 'street soccer' during the mid-'90s. He stressed the positive implications of soccer-specific training for overall motor performance and warned – today we can assume legitimately – against the rise of non-sportive leisure-time behaviour such as video games (Naul, 2001). Roland expressed his passion for the game not only with publications, but also with structural engagement in and around the sport. He was a founding member (1985–1987) and chairperson (2005–2009) of the Commission of Football in the German Society of Sport Science (dvs) as well as member of the school soccer commission of the German Football Association (DFB, 2006–2008). He controversially enriched the debate about soccer education in school and did not hesitate to bring up painful subjects when it was necessary in his opinion, and he did not consider himself as too good to go with the time and its changes. Although I know that his work in the field of soccer is likely to be finished, it is not difficult to still see his commitment to the sport. And with knowing Roland well enough to say that, I am sure he will remain true to soccer.

References

Ichimura, S., & Naul, R. (1991). Cross-Cultural assessments and attributions to female soccer: Japan and West Germany. In J. Standeven, K. Hardman, & D. Fisher (Eds.), *Sport for all. Into the 90s; comparative physical education and sport*; [the Seventh International Society on Comparative Physical Education and Sport Conference was held in the United Kingdom from July 8th-14th, 1990] (pp. 212–220). Aachen: Meyer & Meyer.

Kempe, M., & Memmert, D. (2018). "Good, better, creative": The influence of creativity on goal scoring in elite soccer. *Journal of Sports Sciences, 36*(21), pp. 2419–2423.

Lottermann, S., Laudenklos, P., Friedrich, A., Metaxas, I., Tritschoks, J., Ferrauti, A., & Weber, K. (2003). Technikdiagnostik und Techniktraining im Jugendfußball. In G. Neumann (Ed.),

Fußball vor der WM 2006: Spannungsbogen zwischen Wissenschaft und Organisation pp. 91-105. Köln: Sport und Buch Strauß.

Memmert, D., Baker, J., & Bertsch, C. (2010). Play and practice in the development of sport-specific creativity in team ball sports. *High Ability Studies, 21*(1), 3–18.

Naul, R. (1986). Fußballspielen in der Schule. *Sportpädagogik, 10*(5), 14–25.

Naul, R. (1989). Sportwissenschaftliche Analysen zum Frauenfußball. In R. Naul & W. Schmidt (Eds.), *Beiträge und Analysen zum Fußballsport 2: Referate zu den Tagungen der dvs-Kommission Fussball 1987 und 1988 in Duisburg, Beiträge und Analysen zum Fußballsport* (Vol. 40, pp. 38–60). Clausthal-Zellerfeld: Deutsche Vereinigung für Sportwissenschaften.

Naul, R. (2001). Fußballsport im Freizeit- und Leistungsprofil von Jugendlichen in Europa. In D. Teipel, R. Kemper, & D. Heinemann (Eds.), *Nachwuchsförderung im Fußball. 15. Jahrestagung der dvs-Kommission Fußball vom 17.-19.11.1999 in Jena: Vol. 118* (1st ed., pp. 147–164). Hamburg: Czwalina.

Naul, R. (2005). *Der Deutsche Fußball-Fitness-Test.* In BDFL (Ed.), *Internationaler Trainer-Kongress 18. bis 20. Juli 2005.* Düsseldorf „Aktuelle Trainingsansätze zur. Umsetzung einer modernen Spielphilosophie (pp. 62–70). Frankfurt/M.: BDFL.

Naul, R. (2006). Der Auf- und Abstieg des Fußballspiels an Schulen im Spiegel der Lehrpläne von Nordrhein-Westfalen. In M. Raab, A. Arnold, K. Gärtner, J. Köppen, C. Lempertz, & N. Tielemann, et al. (Eds.), Vol. 2. *Schriftenreihe Human Performance and Sport, Tagungsband* (pp. 61–64). Flensburg: Flensburg University Press.

Naul, R. (2007). Talentförderung im Mädchenfußball: Wege und Probleme. In G. Gadawietz & U. Kraus (Eds.), *Die Zukunft des Fußballs ist weiblich. Beiträge zum Frauen- und Mädchenfußball;* [1. Frauen- und Mädchenfußball-Kongress in Köln, 7–9. April 2005] (pp. 90–107). Aachen: Meyer & Meyer.

Naul, R. (2010). Der Fußballsport in Schule und Hochschule – zwischen Bildung und Ausbildung. In O. Höner, R. Schreiner & F. Schultz (Eds.), *Aus- und Fortbildungskonzepte im Fußball* (pp. 15–31). Hamburg: Czwalina.

Naul, R., Brüggemann, D., & Völz, C. (2008a). Talented players need physical and mental challenges: Part 1: Optimizing performance with total fitness. *Success in Soccer, 11*(4), 28–33.

Naul, R., Brüggemann, D., & Völz, C. (2008b). Talented players need physical and mental challenges: Part 2: Soccer fitness: Laying the foundation for technique training. *Success in Soccer, 11*(5), 16–19.

Naul, R., Hoffmann, D., Altenbeck, B., Brüggemann, D., Völz, C., & Harlemeier, M. (2008). Talented players need physical and mental challenges: Part 3: Psychology: Assessing youth players' mental skills. *Success in Soccer, 11*(6), 30–38.

Naul, R., Richter, C., Hoffmann, D., Altenbeck, B., Brüggemann, D., & Fischer, B., et al. (2006). Fußball-Fitness und Psycho-Mentale Persönlichkeitsmerkmale bei DJunioren im Kader des Fußball- und Leichtathletik-Verbandes Westfalen (FLVW). In K. Weber (Ed.), *Wissenschaftlicher Transfer für die Praxis der Sportspiele. Gemeinsames Symposium der dvs-Kommissionen Sportspiele, Fußball und Tennis vom 18. - 20. November 2004 an der Deutschen Sporthochschule Köln: Vol. 9*. Wissenschaftliche Berichte und Materialien/ Bundesinstitut für Sportwissenschaft (1st ed., pp. 132–143). Köln: Sportverl. Strauß.

Naul, R., Schmagold, J., & Wadle, H.-J. (1987). Zum Studium des Faches Fußball an bundesdeutschen Hochschulen im Rahmen der Sportlehrerausbildung. In R. Naul, G. Tiegel, & W. Schmidt (Eds.), *Beiträge und Analysen zum Fußballsport: Vol. 40*. DVSProtokolle (pp. 153–162). Clausthal-Zellerfeld: DVS.

Niehues, D., & Naul, R. (2017). *"Fußball-Fitness und psycho-mentale Eigenschaften: Vom Talent zum Profi.Ð* 1st Sino-German Forum of Youth Football Research and Development, Tongji-Universität, Shanghai, China, 23.11.2017.

PART TWO:

GENERAL EDUCATION AND SPORT EDUCATION

7 The morphology of the city: Traditions of sports and social work

Niels Rosendal Jensen & Søren Nagbøl

Note

By means of an example – the Brumleby settlement and the football club Skjold – this chapter thematizes urban planning in Copenhagen in the twentieth century as well as the ways of life within these planning stages. The case makes up the first part of the chapter and serves as its empirical base, thus including a residential area, an athletic association as well as a social community, characterized by its considerable political importance for a community located in a capital city. Therefore, the first part is primarily descriptive.

The second part is an analytical approach in which the authors present various viewpoints and approaches to the case, drawing upon social pedagogy and political traditions (self-management and civil society). Various concepts and methods are used and transformed in order to show how social staging influences formation, civilizing and socialization, and how the importance of social staging can be interpreted in the present day. This article hopes to contribute elements to a rethinking of the relationship between sports and social pedagogy.

7.1 The Case

7.1.1 The housing of the medical association: Present-day Brumleby

In the middle of the eighteenth century, Asian cholera flooded Europe with a strength resembling that of the plague epidemics of the Middle Ages. The city of Copenhagen prepared accordingly, or so it thought: the director of the only poor people's hospital in the city placed an order for 200 coffins to be produced. The first cases of illness were recorded in June of the year 1853, and in the following 4 months, the disease afflicted 7,219 people – of the afflicted individuals, 4,737 people died.

Back then, Copenhagen was a small city surrounded by ramparts. The military did not want to give up their fortifications for the bordering demarcation area. A heavy increase of the populace meant very small and crowded housing. The city's drinking water was, in most places, unsanitary, and personal hygiene as well as sanitation standards was poor. In this disastrous situation, 100 of the city's practicing doctors banded together and formed the Medical Association against the Incidence of Cholera, and they volunteered to relieve their publicly practicing colleagues (Skriver, 1971).

The association focused on the miserable conditions of housing for ordinary people, and it succeeded in getting a free making over of a site at Øster Fælled in order to build healthy and cheap houses for 'the poor classes'.

The members of the association focused on the appalling housing conditions for ordinary citizens, eventually succeeding in persuading the magistrate to surrender a plot at Øster Fælled that could be used to build healthy and affordable housing for the 'working class'.

7.1.2 Brumleby: The cradle of the Football Club Skjold

The Brumleby museum website (http://www.brumlebymuseum.dk/bedre-boliger/) describes the history of the social housing as follows:

> *After the spread of cholera in the working class neighbourhoods of Copenhagen, the direct link between bad housing conditions and instances of illness became clear. The association's housings were established in 1853 outside the ramparts of Copenhagen on the plot at Øster Fælled, and the use of systematic hygiene would eventually hold great significance for the layout of the buildings. The blocks were placed at some distance, allowing for fresh air to course between the houses, and sunny spots as well as fresh drinking water was made readily available. The area was established from November 1883 and onwards, and the Association intended to continue building exemplary houses, thus conducting fundraisers.*
>
> Skriver, 1971, p. 23

The basic notions of the village came from the idea of conducting a healthy life as it was seen from the side of the bourgeoisie (Jørgensen, 1997). This was based on guiding principles of fresh air, hygiene and health – in other words, a good life was made up of these elements. Taking the understanding of social pedagogy that we will return to later into consideration, one could argue that this project aimed at solving the social question by means of health and morale of the poor people and of avoiding the creation of a dangerous or even revolutionary proletariat. Hans Helge Madsen (Madsen, 1979, p. 7) notes: This village or neighbourhood was made to establish a workers' quarter in a class society, and in fact, it has been functional according to the original terms since the beginning. On the one hand, the workers were offered cheaper and better apartments compared to what would usually be available to them, and on the other hand, the residents had to obey very strict control of their life at home. The governing body of the residential area exercised this control, and control was made by the inspector and guards on a daily basis.

The originator of Brumleby, Emil Hornemann, who supervised the residential area as a physician, struggled against the causes of poverty and illness. He saw these to be the outcomes of drunkenness, uncleanliness and immorality, and in particular, he found that the behavioural patterns of housewives had to be regulated in order for the workers to realize their own responsibility for avoiding poverty.

7.1.3 The football club

When the Football Club Skjold was established, it soon came to play the role of a 'stamping ground', both in regard to sports and politics. The lawns between the houses functioned as both playgrounds and playing fields for various ball games. One of the founding fathers of the club, Ove Günzel, remembered:

> *The boys played football at the lawn between the buildings J and K, and instead of using ordinary footballs, they used tennis balls. When playing with the tennis balls, they became good players. The trees made up the goals, and throw-ins took place when the ball passed the curb. The housing regulations limited ball games to a certain age group, that is, at the age of confirmation (around 14 years of age), the players had to stop. However, sometimes someone could not remember when he was confirmed.* Madsen, 1979, p. 148, our translation

Skjold was established on 15 February 1915. At the beginning, the club allowed Brumleby residents only. Over the course of the following years, boys from other streets of the surrounding areas begged to take part. During the years of great unemployment in the 1920s, the unemployed youths of the surrounding areas of Østerbro came to the commons at Øster Fælled and played football. In those days the slogan sounded: *'We'll meet at the commons.*

7.1.4 From commons to 'the Copenhagen commons'

On a more national level, in 1901 a change of political systems took place, and cabinet responsibility was introduced. Two years later, the Social Democratic Party and its allies won the majority of the seats in the capital, and this majority elected the first Social Democrat lord mayor of Copenhagen. One of the first initiatives the lord mayor Jens Jensen undertook was the establishment of a council to transform the commons into a city park. They looked to Hyde Park in London for inspiration, and the virtues of the English garden which aimed at an outdoor life that marked the mentality of English gentlemen and their sports was seen as ideal. Steen Eiler Rasmussen (1934, p. 304) points out the basic notion of the London park as a frame for a free outdoor life, in which parks should not just be pretty, neither were they to be content with realizing their own purpose. Rather, parks were to have useful values, perhaps as a kind of technical park, or as grounds of sanitation, that is, resembling a large city's waterworks, sewers, and so on. In Europe, parks are often seen as open areas characterized by fresh air, opposites of stuffy residential housings. For the English, however, the most important mission of the public garden is to provide the populace with the means of a healthy outdoor life. The development of modern parks in London is therefore closely linked to the development of English sports; their design is determined by the demands of popular sport.

Here, we take notice of the fact that the cultures of the workers and of sports are united in a new symbolic order – a park characterized by wide lawns invites new forms of activity and cohesion of daily life. On the one hand are sports like football, cricket and other ball games, and on the other hand, there is a central meeting place for the demonstrations on the May 1.

The new Copenhagen commons exerts a considerable influence on the organization of sports in Copenhagen; for example, the establishment of the Danish Workers' Sport (DAI) in 1929 was improved. An organized football culture arises in the wake of the establishment of the new commons, one that is vibrant and lives on even today.

7.1.5 The Madsen family: One connection between Brumleby and Skjold

The Brumleby history books (page 150) show a photograph taken in 1922 in front of the common house 'Kostalden' ('the cowshed') by some of the prominent members of Skjold.

In the top row to the right, we find Hjalmer Madsen. Güntzel told Hans Helge Madsen: 'When we were 11 or 12 years old, Hjalmer Madsen was made the coach of housing block P. Mads Volmer Madsen was born in Brumleby, the son of Maria and shoemaker Hans Madsen. Mads Volmer Madsen, brother of Hjalmer Madsen, was married to Clara Alfride Rådberg. She had a son (outside of marriage) with KajRådberg from Sweden. Together with Clara Alfride Rådberg, Mads Hjalmer Madsen had three sons: Ingolf, Arne and Tom. For decades Kaj was a player in "Frem" and a cashier in "Skjold". The mother, Clara, did the laundry for the entire club. Arne Madsen (1926–2015) was not just an active player in "Skjold" for most of his life, but also a prominent member of the Social Democrats on Østerbro. Arne Madsen played several thousand matches for the club, and Tom Rådberg (Madsens), grandchild, Per Rådberg Nagbøl, born in 1991, has been an active player in "Skjold" since 1995' (Madsen, 1979, p. 150).

It is quite unusual to follow a family that has lived in the same housing for five generations as they have continued to be active not only in the football club, but also in the local Social Democrat Party union. This continuity of connection to the local workers' environment on Østerbro as well as the clubs of the commons still serves as a guiding principle with both an established physical presence as well as being part of the mentality of 'Skjold's work even today. This can be seen when the 'old boys' meet every Thursday at the club to play pool, at the yearly Easter lunch in Kostalden, at the spring party every April in Brumleby, or the 1 May celebrations in Fælledparken, and so on.

The present chairman of Skjold, Jan Sørensen, is a third-generation resident of Brumleby, hailing from a family that has continued to carry the chairman position in Skjold. Jan's granddad, Carlo Sørensen, was chairman from 29 April 1931 until 14 April 1937. Jan's father, Jørgen Sørensen, was chairman from 31 August 1967 until 26 January 1972. Jan Sørensen himself has been chairman since 30 December 2009. (Information in an interview with Jan Sørensen, 10 January2018). In the interview he further stated the following: Traditions from the early days of Skjold are still alive. The most interesting features are as follows: The club is dependent on volunteers to act as coaches; anyone can become a member; an elitist approach to the players is not accepted, which means that strong and weak players are not split up in two teams but are made parts of the same team.

Though some players may not show up on a regular basis, perhaps due to the need to fit football matches into family life, they are neither excluded nor expelled. When good and bad players play together, learning processes arise, and those processes are not dependent on becoming the champion amongst peers. A mixed team tells every single player to accept and help the players who are not yet masters of the game. The good player has to find room and understanding for other players and thus contributes to creating a game with many variations. Voluntary work creates more opportunities for meeting popular interests and needs. Further, some activities are not aimed at something predefined, but rather resemble our own activities as kids playing football at playgrounds or in the streets. Skjold is not a club for the elite and is not able to match elitist demands. However, the club can do what an elite club cannot do, that is include those who have no other place to live out their dream of playing football. This appeals to boys who are different, and the club and coaches are usually patient in such cases. These boys can participate as they are, and this does not necessarily imply that they cannot become good players in the end. On the contrary, over the past decades, Skjold has produced many gifted players – for example, Pierre Emil Højbjerg (Bayern München and Southampton FC) as well as other gifted players, enough to man a premier league team or even a national team.

The mentality of Skjold emphasizes the social aspect, for example, respect for other people. For example, a former player showed up some years ago and asked to become a coach; he had been sentenced to prison terms ranging from three to five years for the distribution of drugs. The club's response was clear: If you do your job as a coach, that is okay. Attention is focused on his daily work and his future, not his past. His voluntary work has been nearly perfect, and he is currently in a stable relationship with his family and enjoys earning wages. However, despite these positive experiences over eight years, he was not allowed to train a group of pupils. He had no criminal record after ten years. He was accepted and respected as a couch. Though different compared to other members of the club, he was welcomed and not stigmatized. 'The stranger' soon became a peer.

Moreover, the club has had a number of offers for blind people and for young people of various groups, for example, a female football team for lesbian women only. In short, the club accepts and respects various communities and inter-human differences.

7.2 Analytical approaches

This case reveals certain openness related to 'others' or even to problematic situations in instances where sports and social pedagogy are connected. Further, it shows a will to include these 'others' in a community based on sports.

Roland Naul shared this viewpoint in 2007, when he gave a lecture on the tenth anniversary of the Danish Association of Social Pedagogy (Naul, 2007). Amongst other things, Naul epitomized several clues for furthering the relationship between sports pedagogy and social pedagogy: 'As my statements [. . .] have hopefully shown, there existed and exists a considerable historical heritage between these two pedagogical disciplines in the educational sciences. To re-excavate and re-adjust this heritage today, at the beginning of the 21st century, might not only prove fruitful for the two disciplines. In the future, such efforts will become necessary when taking the social and pedagogical transformative processes into account as they make themselves felt in our common life world and in the social spaces of children and young people, as well as their parents, pedagogues, teachers, coaches, and leaders. Let us all, social pedagogues and sports pedagogues, find our way back to our common roots and act jointly by means of "new thinking" on new societal and socio-cultural processes of change, and eventually, we will find a strong and comprehensive response' (Naul, 2007, p. 21, our translation).

Our contribution does not intend to 'solve' the challenges forwarded by Naul. However, we would like to share our reflections regarding the relationship between the two disciplines.

7.2.1 What is social pedagogy?

The meaning of the notion or concept of *social pedagogy* has undergone vast changes over time and been understood in many different ways. There may have been a consensus once upon a time; today, there is no consensus. En passant, we are not interpreting the historical development of social pedagogy as a history of decline nor decay. On the contrary, our approach emphasizes the interrelatedness of society and social pedagogy.

In this context we understand social pedagogy in two senses. In a narrow sense social pedagogy is defined as a profession, whereas in a broad sense it is linked to the 'social question' (in German 'die Arbeiterfrage'). Both understandings are further merged with practice and as such in movement. The narrow sense points rather to methods and ways of working as professionals or as volunteers (e.g., laymen) who aim at helping children or young people in need of help or at risk. The football club is to a very high extent comparable with social pedagogical interventions in the narrow sense. In the broad sense, the whole estate or residential area makes an example of societal interventions aiming at shaping a local solution for working class people in terms of decent housing, social community and sport. Whether we construct social pedagogy in the narrower or the broader sense, we cannot escape the basic idea of education.

In the centre of this conceptual understanding we take notice of the term *education*, which means 'Bildung'. Quoting Korsgaard we sum up: 'The Enlightenment is the beginning of a new era in the history of European ideas. The thinkers of the Enlightenment did not only have enlightenment but also general education at the core of their understanding of becoming a human being. The human being is nothing at the outset, it is formed. Bildung was the essential word in German education and educational theory which became a key notion in an entire pedagogical philosophy. Even though there was no unified theory in modern enlightenment and educational thinking, there was nonetheless a theoretical basis that tied the strands together. This often-overlooked structure is the individual's relation to itself, to society and to the world. This triple relation is the structural basis of what could be considered the trinity of general education theory' *(Korsgaard 2004, pp. 573–574)*. This general theory is shaped differently depending on the understandings from one viewpoint to another. Korsgaard emphasizes four 'isms': National liberalism, Grundtvigianism, Culture radicalism and Socialism that were fighting to 'lay claim on the notion of people. However, a national compromise on the notions, state, nation, people and democracy was reached during WW2, which came to have significant importance on the creation of the Welfare state after the war' *(ibid., p. 590)*. The four 'isms' are interpreting the principle of general education according to the class or the interests each of them represents.

What Korsgaard points out here may be useful in our outline and discussion of social pedagogy. In Denmark, social pedagogy has its own characteristic history, connected with few persons, and with a special type of institution, the folk kindergarten. Kindergartens were established 1871, based primarily on Fröbel's ideas as regards nursing, formation, education and pedagogy. In current years the practice of Danish kindergartens has given Danish social pedagogy a special attention and high reputation noticed by many foreign guests. However, the image of being something unique has at the same time been an obstacle to understanding that Danish social pedagogy has much in common with the social pedagogy of other countries.

7.2.2 Traditions or historical tracks

The existence of different traditions has an outstanding influence. Eriksson and Markström have discovered three different tracks. First and foremost, it is about the continental tradition, which draws on Natorp, whose ideas were related to the German labour movement (a collective comprehension). Subsequently, the attention is drawn to the American tradition that has roots in philanthropy, where Mary Richmond serves as example (a therapeutic and individual comprehension). Finally, as a third one, a pedagogical comprehension on the basis of Dewey and Freire is seen. As Dewey pointed out, education would prove its proficiency in relation to developing democracy (he seems to agree with Natorp in this); Freire stressed that dialogue, raising awareness, emancipation and reflective action are essential elements to transform reality. Since societal problems (e.g., unemployment) create problems for the individual, one must act. But as society does not have the resources, the regional areas must take care of it. Thus, the social pedagogue plays a vital role as facilitator of local initiatives in active collaboration with voluntary organizations. On the one hand this shows the strength of social pedagogy, but on the other, also its weakness, as emancipation is no issue, if the purpose is solely centred at rehabilitation and integration. Thus, we can identify three main streams, namely an adaptation-oriented, a mobilization and a pedagogical approach (Eriksson & Markström, 2000, 2003, pp. 9ff.). One could add a fourth independent tradition, namely the group's role in youth work (Makarenko, 1955; Rasmussen, 1983). An element is recurring in social pedagogy, namely 'the whole person'. This we engage in the next section, which also can be seen as a link between theory and practice.

7.2.3 'The whole person'

Social pedagogy is a carrier of an idea of the human being or the person. To these different interpretations are available, which in turn are based on different concepts of human, society, view of history, scientific school, morality, political beliefs and ideological movements. Some authors see this as a reflection of a particular approach to self and to society. Since Pestalozzi (1746–1827), a feature of social pedagogy has been the tendency to understand the human as a whole. Pestalozzi is considered to be the originator of this perspective, which includes three elements: head, heart and hands. Not to be understood as three parts of a human, but as three angles of a single person – namely, the ideal of the adult human. This implies for one that any social pedagogical effort will require an approach that integrates intellectual, moral and practical education. Personal autonomy, which is a central aim of any pedagogy or education, is the result of an integration that is not achieved by using fragmented efforts.

Just as there is a certain Babylonian confusion about understanding what theoretical social pedagogy is, the same outline recurs when one speaks of pedagogical efforts. According to custom, one is referring to care, help, support, supervision, treatment, mobilization and education. Again, we find different interpretations of what those concepts cover. Pestalozzi was the first one to link help and education together (Horlacher, 2007).

According to the tradition, Natorp found the mobilizing aspect of the labour movement to be an incarnation of the idea of community. Since work is the basis for human identity formation, and it also could lead to liberation and democratic influence, Natorp imagined that work had to be released from alienation (i.e., capitalism) and instead serve human emancipation. The pedagogical task was not to interfere in this liberation struggle, but rather to provide new educational conditions in order to support liberation. Thus, teaching was to translate the challenges of industrial capitalism to a liberation perspective. In Natorp's time this failed and ended up in Hitler's National Socialism and millennial Reich. Under those circumstances social pedagogues gave in to state orders.

The point is that the idea of 'the whole person' appeals to research as well as to professional work, and ties the two logics together.

Based on the above outline we claim that sport may include the three dimensions. Sport has adaptive potential when aiming at performance and how to maximize performance. An increasing commercialism is our proof – in particular related to an elite and a picked football team (star status). Professional football players have unbelievable wage contracts; thereby, the stars obtain high status and excellent economic conditions. To maintain such status some sportsmen are doping themselves. Sport then becomes an instrument or a tool for something else. This takes the very common form of adaption-oriented sport. Another form understands sport as a means of prevention regarding health or criminality or integration of young people at risk. However, sport is at the same time a means of mobilization, in particular in the years of democratic struggle where sport associations became part of mobilizing the peasantry against the Conservative government. This took place from around the 1870s until 1901. Even today sport can mobilize people – and unfortunately not only for good causes as evidenced in hooliganism and extreme nationalism.

Eventually, sport functions as pedagogy – in particular when ascribed a value of its own and on terms of its own. For example, amateurs in sport emphasize well-being, good fellowship, fun and common efforts to reach common goals. A very impressive example is to do with play and fellowship. This last point underlines the importance of the group that may exist in a wide range of shades.

7.2.4 Perspectives from educational sociology and anthropology

Following up the traditions of social pedagogy and political activity, we attempt to transform those concepts and methods by means of assumptions based on educational sociology and anthropology. Within this framework, concepts like staging and analysis of movement contribute to the new thinking of the relationship between sport and social pedagogy. The concepts are movement, time, space, body and patterns of behaviour dependent upon the mentioned concepts (Eichberg, 1987, Nagbøl, 1994).

Siegfried Bernfeld, an Austrian psychoanalyst and educationalist, notes the fact that people while using their faculties create their social world. Humans develop their own experiences by means of interactions and encounters with other individuals. Experiences are at the same time linked to affective relations reaching far beyond a mere cognitive recognition.

Bernfeld rebels against the dominant thinking of pedagogy by emphasizing a developmental perspective of mankind, namely that children grow and develop over many years. This perspective is strongly influenced by biological and societal conditions. Bernfeld states that the child's development is responded to by institutional arrangements in society (Kindergarten, playground, school, etc.). In brief, society responds to the development (Bernfeld, 1967, p. 51).

These responses are 'Erziehung' in Bernfeld's thinking, a concept covering education, personal formation and instruction. The concept draws our attention to the fact that these societal responses have to be taken into account when understanding education, formation and instruction as societal processes and not merely as a system of norms and instructions. This viewpoint may be of importance as regards Naul's appeal for new thinking.

Movements are a key to understanding the mutual exchange and production of the social world (cf. Gebauer & Wulf, 1998). The processes of exchange and production go on at several levels. The personal life of the individual is one side of the coin; the organization of social systems is the other one. When finding the connection between the two sides, we have to examine interactions between actors and their social context (Dietrich, 2002). The processes are expressions of human practice showing that they live in interdependent figurations (Elias, 1970). There are various forms of staging such as nurseries, schools, after-school institutions, playgrounds, parks, toys, private and public spaces, universities, and so on, in short all those coherences that symbolize the societal reaction of man's developmental perspective.

In our case study, more shades of staging are intertwined and thereby influence humans in contexts where they are civilized and socialized. For instance, we have pointed out urban planning, architecture and the local community, Brumleby, which is not solely an issue of architecture but rather a broad perspective on a better life in healthy surroundings, common activities and social coherence.

The process shows how humans live in figurations, including a performative culture of movement enfolded as a new staging characterized by its special dynamic, for example, the football club Skjold realizing how a joint setting of sport and social pedagogy may be capable of including people. We might have taken unnecessary trouble. However, we claim that tradition and innovation contribute to new functions in a modern metropole in the twenty-first century.

References

Bernfeld, S. (1967/1925). Voraussetzungen und Funktion der Erziehung. In *Sisyphos oder die Grenzen der Erziehung.* Frankfurt an Main: Suhrkamp Verlag.

Dietrich, K. (2002). Sport settings in change. In K. Dietrich (Ed.), *Sozialisation and the social change – In movement culture and sport. København*: Institut for Idræt og Ernæring, Københavns Universitet.

Eichberg, H. (1988). *Leistungsräume. Sport als Umweltproblem.* Münster: Lit Verlag.

Elias, N. (1970). *Was ist Sociologie.* München: Juventa Verlag.

Eriksson, L., & Markström, A. M. (2000). *Den svårfångedesocialpedagogiken.* Lund: Studentlitteratur.

Eriksson, L., & Markström, A. M. (2003). InterpretingtheConceptofSocialPedagogy, In A. Gustavsson, H. E. Hermansson, & J. Hämäläinen (Eds.), *Perspectives and theory in social pedagogy* pp. 9–23. Göteborg: Daidalos.

Gebauer, S., & Wulf, C. (1998). *SPIEL – RITUAL – GESTE. Mimetisches Handeln in der sozialen Welt.* Reinbekbei Hamburg: RowohltVerlag.

Horlacher, R. (2007). Die Rolle Pestalozzis in der Konstituerung der Sozialpädagogik als wissenschaftliche Disziplin in der modernenGesellschaft. In B. Dollinger, et al. (Eds.), *Die*

Sozialpädagogische Erziehung des Bürgers: Entwürfe zur Konstitution der modernen Gesellschaft (pp. 159–173). Wiesbaden: VS VerlagfürSozialwissenschaften.

Jørgensen, P. (1997). *Ro, Renlighed og Regelmæssighed. Dansk Idrætsforbund og Sportens gennembrud ca. 1896–1918.* Odense: Odense Universitetsforlag.

Korsgaard, O. (2004). *Kampen om folket: et dannelsesperspektiv på dansk historie gennem 500 år.* København: Gyldendal.

Madsen, H. H. (1979). *Brumleby Historiebog. Lægeforeningens Boliger på Østerbro.* Nationalmuseet. Andelstrykkeriet i Odense.

Makarenko, A. M. (1955). *Vejen til livet: en fortælling om opdragelse.* København: Forlaget Mega.

Nagbøl, S. (1994). Helgoland on Amager. In *International Review for the Sociology of Sport* H. Eichberg (Ed.), *Special issue narrative sociology.* Munich: R. Oldenburg Verlag.

Naul, R. (2007). Socialpædagogiske udfordringer, sport og idræt i børn og unges institutionaliserede hverdagsliv. *Tidsskrift for Socialpædagogik, 20,* 11–22.

Rasmussen, H. (1983). *Gruppeopdragelse.* København: Forlaget Tiden.

Rasmussen, S. E. (1934).*London.* København: Nordisk Forlag.

Rasmussen, S. E. (1994).*København. Et bysamfunds særpræg og udvikling gennem tiderne.* København: C.E.C Gads Forlag.

Skriver, P. E. (1971). *Bevaringen of laegeforeningens boliger. Betaenkning afgivet af et udvalg nedsat of Danske Arkitekters Landsforbund.* Kobenhavn: Arkitektens Forlag.

Toft J. (1993). *Fodbold mellem myter og kilder I: Idrætshistorie og Livshistorie. Idrætshistorisk Årbog 1993.* Odense: Odense Universitetsforlag.

8 Physical activity and sports in German all-day schools: Status quo and challenges

Nils Neuber

Growing up of children and adolescents in Europe has changed severely in the last 10–15 years. Against the backdrop of social modernization processes, it has become more insecure and contradictory, whereas modern 'multi-option societies' offer *varieties of opportunities of action* to young people in the sociocultural sector. What is more, it is an increasing dynamization and digitalization of everyday life. All in all, this has led to a *detraditionalization of CVs* with a simultaneously increasing responsibility for the private way of life (Neuber & Salomon, 2015). While the majority of young people succeed in making positive use of the altered life conditions, a growing number fails because of modern challenges. In Germany, almost one out of three adolescents comes from a family home, which is threatened by poverty, where the parents have no job or insufficient qualifications (BMFSFJ, 2013, p. 54). Nevertheless, unemployed youth in Germany form only 6 per cent – in comparison, in Italy there are 33 per cent young people without jobs, in Spain 34 per cent and in Greece 45 per cent (Statista, 2018).

As a reaction to these developments, all EU members have increased their efforts in education. There is no doubt that education is key to employment and prosperity. In view of this fact, many countries have intensified their collaboration in municipal networks (Süßenbach & Klaus, 2015). Sports thereby gains a special importance. In 2002, Great Britain employed a new 'Physical Education and Sport Strategy', which aims to improve the connection between PE, school sport and club links. For a considerable time, the Netherlands has focused on sports-based municipal networking as well (Naul, Dreiskämper & L'Hoir, 2015). School sports clubs in France have similar aims (Brandl-Bredenbeck & Grimminger, 2015). In this regard, Germany has tried to expand its all-day schools. Nowadays German children and adolescents spend distinctly more time in school than before. Whereas in the school year 2002/03 less than 5,000 schools were all-day schools; in the school year 2013/14 there were already more and 16,000 which is about half of all schools (KMK, 2015). Hence, within a few years, the all-day school has become an integral part of the German education system (Coelen & Stecher, 2014, p. 5).

This has immediate *effects on the leisure behaviour* of children and adolescents, especially on their exercise, play and sports activities. Admittedly, there is no proof that adolescents do less sports due to all-day schools; beyond question is, however, a *change in sports commitment*. Cooperating with extracurricular partners, the all-day school's job is now to organize suitable sports activities for children and adolescents. The question arises, how the exercise activities should be arranged. It is agreed upon offering more than just a smooth extension of club offers or just a doubling of curricular PE (Naul, 2006, S. 21). Indeed, there is a new *type of offer in the intersection of sports and social pedagogy*, which is to be placed between PE and club practice, between formal and informal learning opportunities (Neuber, 2008). Often this is called a new, third column of children and adolescents sports (fig. 8.1). Characteristic of all-day school activities is the missing pressure for good grades, little competitive behaviour, often inter-year groups and the instruction by educational professionals, for example, trainers and coaches (Derecik, Kaufmann & Neuber, 2013, pp. 37–38). These special circumstances are a major challenge for the all-day school staff since classical concepts of school and club sports can only be applied partly.

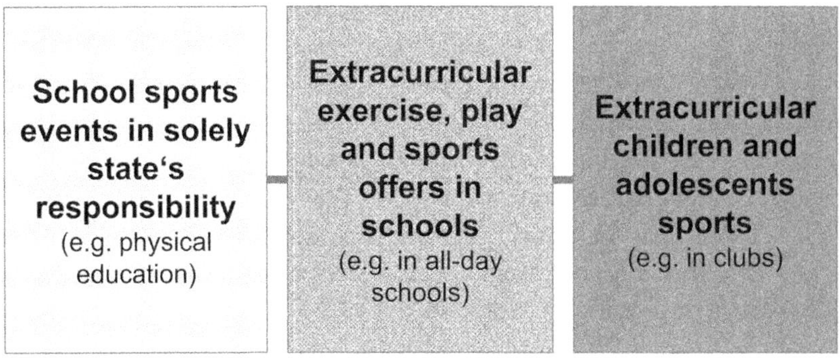

Figure 8.1: Three-column model of children and adolescents education in sports (Pack & Bockhorst, 2011, p. 169)

Physical activities in all-day schools also bring up new tasks for *(sports) science*. These new types of offers have to be described normatively and captured empirically. While the normative depiction began quite early (e.g., Naul, 2005; Laging & Hildebrandt-Stramann, 2006; Neuber, 2008), empirical findings regarding sports in all-day schools are meagre. Amongst a few general statistics, which mainly focus on federal, scientific research for the development of all-day schools (StEG studies), there are no systematic, nationwide surveys regarding the exercise, play and sports activities in all-day schools. This is why it is hardly possible to make any verified statements concerning goals, staging and impacts of exercise offers; likewise, the perspective of children and adolescents has only been examined rudimentarily (Neuber, Kaufmann & Salomon, 2015, pp. 423–426). A systematic reporting of education in all-day sports is alien to our subject. In this respect, the following compilation of selected empirical results has to be read carefully. Still, some tendencies can be derived, which hint at challenges of all-day sports development in Germany.

8.1 Cooperation of schools and clubs at all-day schools

The majority of all-day schools in Germany offers open models. For a comprehensive introduction of bound all-day schools, which would be compulsory for all pupils, the requirements are missing – both constructional, personal and financial (Rauschenbach, Arnoldt, Steiner & Stolz, 2012, p. 12). It is therefore understandable that the majority of *all-day schools* collaborates with extracurricular partners. When grouping these partners, around 52 per cent of schools cooperate with child and youth service partners (e.g., youth organizations, welfare, youth centres, youth social services), 56 per cent cooperate with cultural-religious partners (music and art schools, libraries, church services) and about 71 per cent of schools cooperate with a sports-based partner, whereas all-day schools claim the highest rate of 85.9 per cent (StEG, 2012, pp. 30–32).[22] Child and

22 The categories might be indistinct, since common good–oriented child and youth sports is part of child and youth services as a private agency.

youth sports is thereby the most common partner for cooperation throughout all-day schools of all sorts (fig. 8.2).

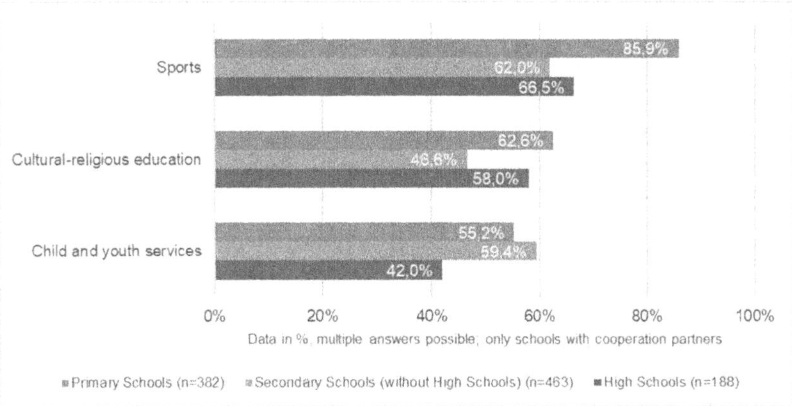

Figure 8.2: Cooperation partners of all-day schools differentiated by cooperation partners and school forms (mod. after StEG, 2012, pp. 30–32)

From the sports club's perspective, this result is less positive. According to the sports development report of 2011–2012, only 17.5 per cent of Deutscher Olympischer Sportbund (DOSB) – organized clubs cooperate with all-day schools. Larger clubs with more than 2,500 members cooperate about seven times more often with all-day schools than smaller clubs with up to 100 members (Breuer, Feiler, Wicker, 2013, p. 2). When looking at the federal states, the cooperation quotas of sports clubs do not differ severely. However, in relation to the existing all-day schools, some serious differences become visible. In Saxony for example, 96.5 per cent of schools are all-day schools, and 24.3 per cent of sports clubs collaborate with those. In Baden-Wuerttemberg, it is only 26.7 per cent of schools, but already 17.2 per cent of sports clubs (Breuer, Feiler & Wicker, 2013, p. 5). Other surveys more or less underline these quotas of cooperation (e.g., Thieme, 2014).

Regarding the *goals of cooperation*, there is a rather consistent opinion. Schools aim at sports-related goals like promotion of physical activity or the experience of new sports. Also the cognitive-dominated school routine should be rhythmized through exercise, play and sports offers (Thieme, 2014, p. 13). At first, the clubs aimed for the acquisition of new members and talent funding (e.g., Derecik, Habbishaw, Schulz-Algie & Stoll, 2012, p. 43). By now, this has changed: aside from the acquisition of new members, pedagogical goals can be identified, such as health promotion or the integration of immigrant children (Thieme, 2013; Süßenbach & Geis, 2014; Naul & Wick, 2015). There are even political goals, such as the positioning of the club in the municipal educational landscape and increase in importance (Süßenbach & Geis, 2014, pp. 28–30). It is remarkable that the individual support of children and adolescents – a key goal of all-day schools – is not of special importance, neither for schools nor sports clubs.

The satisfaction with the cooperation is quite high. Eighty-eight per cent of the schools and 65 per cent of the sports clubs in Hesse find their goals in collaboration in all-day schools have been achieved (Derecik et al., 2012, p. 43). In North-Rhine Westphalia, 73.2 per cent of school representatives are satisfied regarding their sports offers; as the providers of all-day schools with 72.6 per cent. Sports clubs and coordination offices of city and county sports associations only feel 40.0 per cent positive about it (SpOGATA, 2012). There seems to be a need for optimization, especially on behalf of the organized sport. Reasons for dissatisfaction are 'insufficient staff' (13.3%), 'other circumstances' (10.3%) and the 'budget' (9.9%) (SpOGATA, 2012, pp. 17–18). Other reasons, mentioned in a study from Lower Saxony, are 'no communication at eye level' (32.7%), 'great turnover in pupils' (30.8%) as well as 'dissatisfaction with the school people in charge' (22.4%) (Süßenbach & Geis, 2014, p. 33).

8.2 Exercise, play and sports offers at all-day schools

Sports activities are offered at almost every German all-day school, which makes them a nationwide concept (Neuber, Kaufmann & Salomon, 2015, pp. 430–431). However, the organizational forms vary from one school to another: exercise offers are self-organized, for example, during lunch break, as well as instructed, for example, in project groups.

There is little research regarding the *open free-time offers in sports.* A qualitative approach regarding informal learning on playgrounds of all-day schools by Derecik (2011) differentiates between rooms of exercise, rooms of games, sports rooms, activity islands and islands of retreat and communication, which pupils can use informally. It must be assumed that exercise, play and sports offers are popular especially during breaks and recesses at all-day schools. This is supported by surveys regarding the needs of children and adolescents at all-day schools, who unanimously wish for more free and unsupervised time (e.g., Kolbe, Reh, Idel, Fritzsche & Rabenstein, 2009, pp. 28–31).

With regard to the instructed offers, one can differentiate between specific sports and activities that span sports types (overlapping sports). Depending on the study, the numbers vary between 52 and 71 per cent for sport-specific and between 25 and 37 per cent for activities that span sports types. In total, this means a relation of 2:1 (Naul, Wick & Geis, 2015, p. 40). Plus, one can state that the higher the direct participation of sports clubs, the higher is the area of sport-specific offers (Naul, Wick & Geis, 2015, p. 43). For a deeper differentiation of offers, multiple studies can be considered (fig. 8.3): ball sports are in the lead in sport-specific offers in all samples, for example, 21.4 per cent in North-Rhine Westphalia (SpOGATA, 2012, p. 14), 29.1 per cent in Lower Saxony (Süßenbach & Geis, 2014, p. 21) and 33 per cent in Hesse (Derecik et al., 2012, p. 42). Within ball sports – not surprisingly – football is the most popular sport, which makes it the all-day sport number one.

The remaining results are inconclusive. Court sports take the second place in both Hesse (14%) (Derecik et al., 2012, p. 42) and in sports clubs in Lower Saxony (12.5%) (Süßenbach & Geis, 2014, p. 21). A survey in schools in Lower Saxony, however, only states a share of 8.8 per cent for court sports (Süßenbach & Geis, 2013, p. 11). Similarly confusing are findings in dancing. Polls in Lower Saxony schools find a share of 11.9 per cent, surveys in clubs only 5.4 per cent. Derecik et al. (2012) diagnose 11 per cent for Hesse, Naul, Wick and Geis (2015) and 5.9 per cent for North-Rhine Westphalia. Supply numbers in gymnastics vary between 4.3 per cent (school study Lower Saxony) and 9.9 per cent (club study Lower Saxony). In swimming, you find numbers between 2.4 per cent (school study Lower Saxony), 4.0 per cent (Hesse), 4.9 per cent (North-Rhine Westphalia) and 7.4 per cent (club study Lower Saxony). Once again it becomes clear, that, due to heterogeneous findings, secure statements with regard to all-day offers are only partly possible.

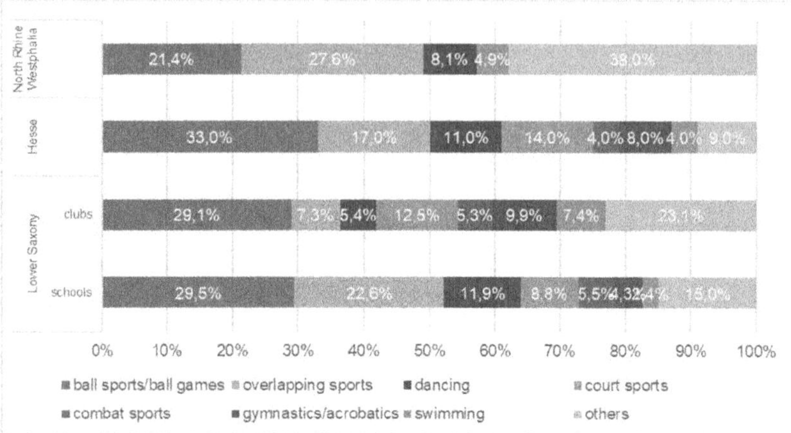

Figure 8.3: Overview of the most frequent sports and fields of exercise in selected all-day school studies (cf. Süßenbach & Geis, 2014; Derecik et al., 2012; Geis, Wick & Naul, 2015)

Regarding the *activities that span across sports types,* it is even harder to make reliable statements, since the categorizations are even more inconsistent. A survey in Hesse sees 'general support of exercise' (9%) and 'psychomotorics' (8%) similar to 'acrobatics' (8%) and 'dancing' (11%) (Derecik et al., 2012, p. 42). Studies in Lower Saxony, however, differentiate between 'support of exercise/mass sports' (24.9% resp. 7.3%), 'dance sports' (11.9% resp. 5.4%), 'health sports' (1.9% resp. 1.6%) as well as 'adventure sports' (1.0 resp. 1.1%) (Süßenbach & Geis, 2014, pp. 21–22). It is worth noticing that the share of sport-overlapping offers at primary and special-needs schools is almost twice as high (13.7% resp. 14.8%) as offers at secondary schools (Süßenbach & Geis, 2014, pp. 20–21). Moreover, there is no advice regarding the methodological *staging of sports offers at all-day schools.* The way of teaching, which greatly influences the organization of the new type 'sports in all-day schools', has not been examined yet.

8.3 Pedagogical professionals in all-day schools

At bound all-day schools, teachers take up the main part of exercise offers. At open all-day schools, however, which account for the majority of schools, mainly other professionals are called into action. The so-called *wider pedagogical staff* consists of educators (31%), social workers (11%) and professionals with a similar pedagogical education (25%) (Coelen & Rother, 2014, p. 113). This clearly highlights the idea of so-called *multi-professional teams* in all-day schools. Compared to other providers, sports staff more often work on a voluntary basis or take it as a second job (Züchner & Rauschenbach, 2011). At all-day schools in Lower Saxony, voluntary and adjunct trainers constitute the majority of the entire staff as well. Nevertheless, there is a tendency towards paid, secondary jobs. The NRW study only finds 14.5 per cent voluntary, but 56.1 per cent adjunct and 29.4 per cent full-time workers (Gerhardt, Wick & Naul, 2015, p. 78). The years after that will show, if the *tendency towards professionalization* continues.

The question of the staff's qualification is still unanswered (fig. 8.4). Still, the highest importance is attached to trainers across all studies: 44.3 per cent of staff at primary schools, 31.2 per cent at secondary schools and 37.5 per cent at high schools have a trainer license in Lower Saxony (Süßenbach & Geis, 2013, p. 15). In Rhineland-Palatinate, 35.1 per cent of the investigated primary school staff own a C-license, 17.0 per cent a B-license and 27.0 per cent resp. 12.7 per cent at middle schools (Thieme, 2012). Pedagogical professionals without sports-related qualifications make up 12.7 per cent of schools in Lower Saxony and 18.0 per cent of primary schools in North-Rhine Westphalia (Süßenbach & Geis, 2013; Gerhardt, Wick & Naul, 2015). Regarding the staff's qualification, the future will also tell which tendency succeeds: less qualified, inexpensive assistants or higher qualified professionals. Increasing institutionalization, however, also requires child and youth sports to further develop the qualifications of its all-day staff systematically.

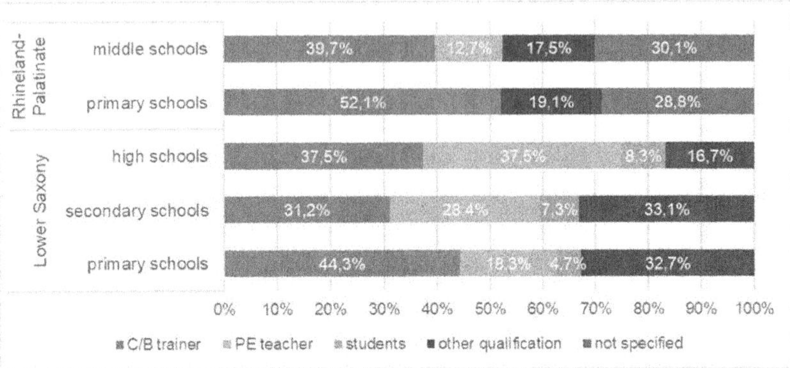

Figure 8.4: Qualifications of providers of exercise, play and sports offers in selected all-day school studies (cf. Süßenbach & Geis, 2013; Thieme, 2012)

8.4 Children and adolescents in all-day schools

The all-day school in Germany is aimed at the individual and social development of its pupils as well as at workplace and family policy issues (Stötzel & Wagener, 2014). The question arises, however, if the *perspective of children and adolescents* is still considered. Those few studies examining the perspective of young people prove the ambivalent relation of them to all-day schools. While offers in all-day schools are seen as assets, girls and boys universally wish for shorter school days. Furthermore, they wish for more unsupervised hours to interact with their peers (Salisch, Kanevski, Philipp, Schmalfeld & Sacher, 2010, pp. 215–219). At least, present studies document a high and very high satisfaction in young people with the exercise and sports offers. They predominantly have good relations with their sports instructors, feel well and fairly treated and receive help upon request. However, regarding possibilities of participation they are distinctly more cautious (Heim, Prohl & Bob, 2013, p. 153).

One all-day school study examines the participation processes of children in exercise, play and sports offers (Derecik, Kaufmann & Neuber, 2013). Unsuccessful participation processes are absolutely daily fare. Moreover, everyday situations offer multiple possibilities of participation, for example, in discussion groups, during setup and deconstruction, and during games, breaks and recesses. Besides specific circumstances, the study identified six 'places' of participation in all-day schools (fig. 8.5). However, the children's range of action between heteronomy, participation and self-determination has proved itself to be complex (Messmer, 1995). Finally, to achieve a gradual introduction to participation processes for children, an education to and through participation is necessary. Besides the establishment of a participation culture at all-day schools, specific suggestions on a didactic-methodological level are required (Derecik, Kaufmann & Neuber, 2013, pp. 221–225).

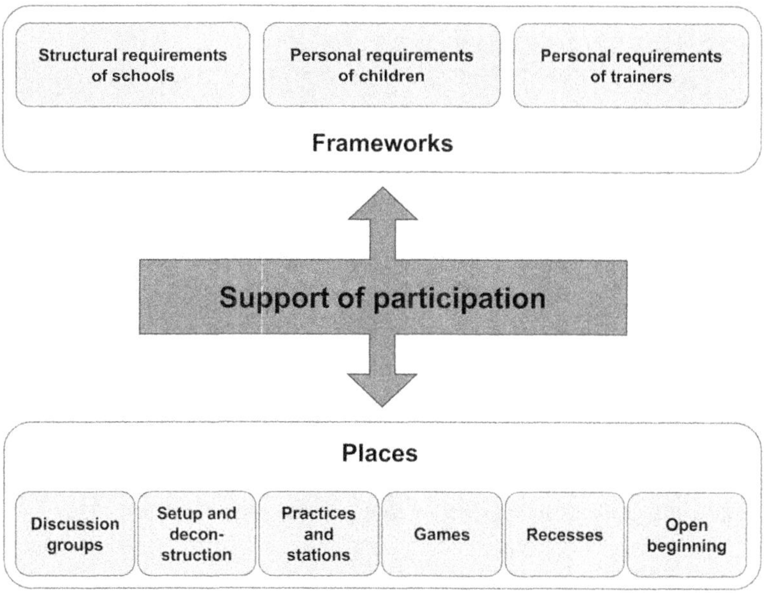

Figure 8.5: Frameworks and places of participation support in exercise, play and sports offers at all-day schools (Derecik, Kaufmann & Neuber, 2013, p. 217)

8.5 Impacts of all-day schools

There had been high expectations regarding the impacts when all-day schools were implemented. Indeed, there are some first, positive findings. Concerning the families, there are slight but positive effects with regards to both the *family atmosphere* and the parents' possibility to pursue their employment (Züchner, 2011). In terms of *educational justice*, a positive effect of learning time and homework assistance on grades at all-day schools (Rother & Stötzel, 2014, p. 140) for migrant pupils was found. In general, children who participate regularly at all-day school offers, show a more positive development of their learning performance compared to others and are less likely to stay down a year. Impacts with regards to social learning were identified as well, for instance a lower tendency towards violence. Taken as a whole, impacts are the more evident, the more intensely children and adolescents participate and the better the quality of all-day school offers (Kielblock, Fraij, Hopf, Dippelhofer & Stecher, 2014).

In general studies, impacts of exercise offers on the learning performance of children have not been investigated yet. Sports-related studies are limited to overlapping aspects of development support through exercise, play and sports. Thus, all-day schools have been asked for their reasoning of implementing exercise in the school routine. Most schools agreed on compensatory reasons (deficit compensation/health promotion), followed by arguments of development support (development support through exercise) and a sports-oriented support in terms of an education to sports (Laging & Stobbe, 2011, pp. 212–213). A longitudinal study by Heemsoth, Tietjens, Naul and Dreiskämper (2015) on the impact of various all-day school sports offers could prove certain effects in regard to BMI, motor performance abilities and self-concept. However, the methodological alignment of the study does not allow generalizations. In this sense, empirical results regarding *effects of locomotor* support in all-day schools are still pending.

8.6 Challenges for the sport in all-day schools

Within only a few years, the all-day school has fundamentally changed the German educational landscape. Exercise, play and sports activities are amongst the popular and most common offers. Between school and club sports, the *all-day school* sport has been established as an innovative concept at schools across the whole country. While schools have been able to adjust gradually to the altered circumstances, extracurricular educational providers have been confronted abruptly with those new challenges. Considering that the common good-oriented children and adolescents sport is by far the largest provider of extracurricular educational offers it becomes clear how serious the challenges are for the system of 'sports', particularly since the sports clubs are largely organized voluntarily and on a second-job basis. However, the expansion of all-day schools offers developmental opportunities especially for children and adolescents sports. Based on the current findings, a few key *challenges for the sports at all-day schools* can be derived:

1. **Clarifying goals – individual support at all-day schools**: Individual support of all children and adolescents has been a key goal of all-day schools right from the beginning. Within the context of exercise, play and sports offers, however, this is neither a central goal for schools nor for sports clubs. This is all the more unfortunate, since from a subject-didactic perspective, there are promising starting points for individual support: the support of exercise, play and sports can be accomplished by various concepts regarding the support of basic motor skills in sports remediation; also, the established talent support programs promise good approaches to support athletic talents. With regard to the support through exercise, play and sports there are both concepts for development support, for example, regarding self-concept, and others for the support of learning performance in schools, for example, concerning executive functions (Pfitzner & Neuber, 2012).

2. **Developing quality – staging of offers:** There are multiple suggestions for the methodological staging of all-day school offers in sports. Considering current papers on task culture, open forms of internal differentiation should be preferred over closed ones. This means more assigning of tasks and less giving of instructions. A higher degree of self-sufficiency of children and adolescents will lead to an altered 'pupil role' ('young people as protagonists of their learning processes') as well as to an altered 'teacher role' ('teachers as learning companions'). The deliberate staging of learning processes at all-day schools can also address specific target groups, for example, regarding gender- or health-specific issues as well as topics of inclusion or aid to refugees (Pfitzner & Neuber, 2012).

3. **Enabling participation – needs of children and adolescents:** It is still uncertain, how far all-day schools satisfy the needs of children and adolescents. Present findings indicate poor possibilities of participation and little space for interest-based leisure activities. In the light of an increasing institutionalization, this is very alarming. Children and youth sports with its high identification potential, however, offers promising approaches for participation support, which underlines its youth league character as well. Extreme positions of both heteronomy and self-determination are useless in this regard. Democratic behaviour in sports can only be learned through practice. This requires sensitive pedagogical support (Derecik, Kaufmann & Neuber, 2013).

4. **Expanding qualifications – multi-professional teams at all-day schools:** All-day schools rely on multi-professional teams consisting of school teachers, educators, social workers and further pedagogical staff. For voluntary and second-job sports offers, the trainer C-license is the first key qualification. It remains uncertain, to what extent this will be sufficient in the future. The tendency towards stronger professionalization calls for adequate advanced training development, including extensions in organized sports in the license system as well as cooperation with other advanced training providers, for example, universities. Those on their part are required to develop new paths of qualifications (Binnenbruck, Krüger & Neuber, 2015). In addition, physical education teachers are asked to shape school sport development at all-day schools even more than before (Neuber & Jordens, 2012).

5. **Developing cooperation – sport as a partner of all-day schools:** Sport is the most frequent cooperation partner of all-day schools; however, not all sports clubs cooperate equally with all-day schools. The clarification of the own profile is essential as well as an agreement on goals and contents of the collaboration. This includes the creation of eye-level communication. Bigger clubs with full-time staff currently succeed better in this communication than smaller ones. This requires sports associations to develop adequate supporting offers. Simultaneously, municipal education players should seize on the quantitative and qualitative importance of sports offers at all-day schools even more than before. Only a systematic interlocking of providers and offers can ensure an educational landscape suitable for its audience permanently (Neuber & Golenia, 2015).

6. **Securing impacts – effects of sports offers at all-day schools**: Last but not least the impacts of sports offers at all-day schools should not be left to chance. Especially since it is a new type of offer in children and adolescents sports, efforts regarding quality development at all-day schools should be intensified. Aside from a definition of standards, which ought to be evaluated regularly, a systematic education report concerning sports at all-day schools should be considered. This can work by implementing sports-related inquiries into the general educational reports as well as through separate studies for the purpose of a sports scientific impact research. Such studies are complex and a lot of work; for school sports only a few papers exist (Neuber, 2009). However, sports offers at all-day schools are too important at sports, educational and youth-political levels to leave their impact to chance.

References

Arnoldt, B., Furthmüller, P., & Steiner, C. (2013). *Ganztagsangebote für Jugendliche. Eine Expertise zum Stellenwert von Ganztagsangeboten für Schüler/innen ab der 9. Klasse.* Last access 30 November 2015: http://www.allianz-fuer-jugend.de/downloads/StEG_Expertise.pdf

Binnenbruck, A., Krüger, M., & Neuber, N. (2015). Wissenstransfer "sport-lernen" – Weiterbildung für Sportlehrkräfte am Institut für Sportwissenschaft (IfS) der Universität Münster. *Sportunterricht, 64*(10), 306–310.

Brandl-Bredenbeck, H.-P., & Grimminger, E. (2015). Frankreich: Ein Systemvergleich. In W. Schmidt, N. Neuber, T. Rauschenbach, H. P. Brandl-Bredenbeck, J. Süßenbach, & C. Breuer (Eds.), *Dritter Deutscher Kinder- und Jugendsportbericht. Kinder- und Jugendsport im Umbruch* (S. 525–547). Schorndorf: Hofmann.

Breuer, C., Feiler, S., & Wicker, P. (2013). *Sportentwicklungsbericht 2011 – Analyse zur Situation der Sportvereine in Deutschland – Schulpolitische Veränderungen und Sportvereinsentwicklung.* Last access 30 November 2015 :http://www.dosb.de/fileadmin/fm-dosb/arbeitsfelder/wiss-ges/Dateien/SEB_Welle_4/SEB_2011_Schulpolitische_Veraenderungen_und_Sportvereinsentwicklung.pdf

Bundesministerium für Familie, Senioren, Frauen und Jugend) (BMFSFJ). (Ed.). (2013). *14. Kinder- und Jugendbericht – Bericht über die Lebenssituation junger Menschen und die Leistungen der Kinder- und Jugendhilfe in Deutschland.* Berlin: BMFSFJ.

Coelen, T., & Rother, P. (2014). Weiteres pädagogisch tätiges Personal an Ganztagsschulen. In T. Coelen & L. Stecher (Eds.), *Die Ganztagsschule. Eine Einführung* (S. 111–126). Weinheim: Beltz Juventa.

Coelen, T., & Stecher, L. (Eds.). (2014). *Die Ganztagsschule. Eine Einführung.* Weinheim: Beltz Juventa.

Derecik, A. (2011). *Der Schulhof als bewegungsorientierter Sozialraum. Eine sportpädagogische Untersuchung zum informellen Lernen in Ganztagsschulen.* Aachen: Meyer & Meyer.

Derecik, A., Habbishaw, J., Schulz-Algie, S., & Stoll, M. (2012). *Sport im Ganztag. Ergebnisse der Initiative "Sportverein plus SchuleĐ.* Frankfurt a. M.: Sportjugend Hessen.

Derecik, A., Kaufmann, N., & Neuber, N. (2013). *Partizipation in der offenen Ganztagsschule – Pädagogische Grundlagen und empirische Befunde zu Bewegungs-, Spiel- und Sportangeboten* (Bildung und Sport, 3.). Wiesbaden: Springer.

Derecik, A., & Laging, R. (2014). Forschungsdesign und Forschungsmethoden der Studie. In R. Hildebrandt-Stramann, R. Laging, & J. Teubner (Eds.), *Bewegung und Sport in der Ganztagsschule – StuBSS: Ergebnisse der qualitativen Studie. Studie zur Entwicklung von Bewegung, Spiel und Sport in der Ganztagsschule (StuBSS)* (S. 84–115). Hohengehren: Schneider.

Gerhardt, A., Wick, U., & Naul, R. (2015). Die individuellen BeSS-Anbieter an Ganztagsgrundschulen. In SpOGATA (Ed.), *Evaluation der Bewegungs-, Spiel- und Sportangebote an Ganztagsschulen in Nordrhein-Westfalen* (S. 75–90). Aachen: Meyer & Meyer.

Heemsoth, S., Tietjens, M., Naul, R., & Dreiskämper, D. (2015). Bewegungsbildung und Gesundheitsförderung in offenen Ganztagsschulen. In SpOGATA (Ed.), *Evaluation der Bewegungs-, Spiel- und Sportangebote an Ganztagsschulen in Nordrhein-Westfalen* (S. 137–164). Aachen: Meyer & Meyer.

Heim, C., Prohl, R., & Bob, A. (2013). Ganztagsschule und Sportverein. In R. Hildebrandt-Stramann, R. Laging, & K. Moegling (Eds.), *Körper, Bewegung und Schule. Teil 1: Theorie, Forschung und Diskussion* (S. 136–156). Immenhausen: Prolog.

Hildebrandt-Stramann, R., Laging, R., & Teubner, J. (2014). *Bewegung und Sport in der Ganztagsschule – StuBBS: Ergebnisse der qualitativen Studie. Studie zur Entwicklung von Bewegung, Spiel und Sport in der Ganztagsschule (StuBSS).* Baltmansweiler: Schneider.

Kielblock, S., Fraij, A., Hopf, A., Dippelhofer, S., & Stecher, L. (2014). Wirkungen von Ganztagsschulen auf Schüler/innen. In T. Coelen & L. Stecher (Eds.), *Die Ganztagsschule. Eine Einführung* (S. 155–171). Weinheim und Basel: Beltz Juventa.

Kolbe, F.-U., Reh, S., Idel, T.-S., Fritzsche, B., & Rabenstein, K. (Eds.). (2009). *Ganztagsschule als symbolische Konstruktion – Fallanalysen zu Legitimationsdiskursen in schultheoretischer Perspektive.* Wiesbaden: VS.

Kultusministerkonferenz (KMK). (2015). *Allgemein bildende Schulen in Ganztagsform in den Ländern in der Bundesrepublik Deutschland – Statistik 2009 bis 2013.* Last access 30 November 2015 :http://www.kmk.org/fileadmin/pdf/Statistik/GTS_2013_Bericht.pdf

Laging, R., & Hildebrandt-Stramann, R. (2006). *Ganztagsschulen bewegungsorientiert entwickeln – Zwischen Schul- und Sportperspektive (Vortrag auf der Tagung der DGfE-Kommission Sportpädagogik am 24.11.2006).* Last access 30 November 2015 :http://www.ganztaegig-lernen.de/media/Material_1/karlsruhe_dgfe.pdf

Laging, R., & Stobbe, C. (2011). Bewegungsaktivitäten als Bildungsangebot in der Ganztagsschule – Ergebnisse einer Befragung aus dem Projekt StuBSS. In M. Krüger & N. Neuber (Eds.), *Bildung im Sport. Beiträge zu einer zeitgemäßen Bildungsdebatte* (S. 201–223). Wiesbaden: Springer.

Messmer, R. (1995). *Partizipation im Unterricht* (Lizentiat). Bern: Universität Bern.

Naul, R. (2005). Bewegung, Spiel und Sport in offenen Ganztagsschulen. *Sportunterricht, 54*(3), 68–72.

Naul, R. (2006). *Sozialpädagogische und sportpädagogische Profession im Ganztag: Qualitätskriterien und Fortbildungsbausteine für Angebote mit Bewegung, Spiel und Sport. Expertise für das BLK-Verbundprojekt "Lernen für den GanzTagĐ.* Essen: Universität Duisburg-Essen.

Naul, R., Dreiskämper, D., & L'Hoir, M. (2015). Niederlande und Großbritannien: Befunde und Interventionen. In W. Schmidt, N. Neuber, T. Rauschenbach, H. P. Brandl-Bredenbeck, J. Süßenbach, & C. Breuer (Eds.), *Dritter Deutscher Kinder- und Jugendsportbericht. Kinder- und Jugendsport im Umbruch* (S. 503–524). Schorndorf: Hofmann.

Naul, R., & Wick, U. (2015). Das Engagement der Sportvereine im offenen Ganztag. In SpOGATA (Ed.), *Evaluation der Bewegungs-, Spiel- und Sportangebote an Ganztagsschulen in Nordrhein-Westfalen* (S. 47–74). Aachen: Meyer & Meyer.

Naul, R., Wick, U., & Geis, S. (2015). Die Bewegungs-, Spiel- und Sportangebote an Ganztagsgrundschulen. In SpOGATA (Ed.), *Evaluation der Bewegungs-, Spiel- und Sportangebote an Ganztagsschulen in Nordrhein-Westfalen* (S. 23–46). Aachen: Meyer & Meyer.

Neuber, N. (2008). *Ganztagsschule – Bewegungs-, Spiel- und Sportangebote in Kooperation von schulischen und außerschulischen Partnern.* In H. Lange & S. Sinning (Eds.), *Handbuch Sportdidaktik* (S. 260–275). Balingen: Spitta.

Neuber, N. (2009). Wirkungsforschung im Schulsport? – Probleme und Möglichkeiten der empirischen Überprüfung normativer Leitideen. In E. Balz (Ed.), *Sollen und Sein in der Sportpädagogik – Beziehungen zwischen Normativem und Empirischem* (Forum Sportpädagogik, 1; S. 11–13). Aachen: Shaker.

Neuber, N., & Golenia M. (2015). Kinder- und Jugendsport als Partner in Bildungsnetzwerken. Grundlagen und Ergebnisse eines Pilotprojekts. In E. Hübner, K. Reinhart (Eds.), *Sport – Geschichte – Pädagogik. Festschritt zum 60. Geburtstag von Michael Krüger* 1. Aufl. (S. 332–347). Hildesheim: Arete Verlag.

Neuber, N., & Jordens, J. (2012). Verschlafen die Sportlehrkräfte den Ganztag? – Zum Wan-del der Sportlehrerrolle in kommunalen Bildungslandschaften. *Sportunterricht, 61*, 291–296.

Neuber, N., Kaufmann, N., & Salomon, S. (2015). Ganztag und Sport. In W. Schmidt, N. Neuber, T. Rauschenbach, H. P. Brandl-Bredenbeck, J. Süßenbach & C. Breuer (Eds.), *Dritter Deutscher Kinder- und Jugendsportbericht. Kinder- und Jugendsport im Umbruch* (S. 416–443). Schorndorf: Hofmann.

Neuber, N., & Salomon, S. (2015). Aufwachsen im Wandel. In W. Schmidt, N. Neuber, T. Rauschenbach, H. P. Brandl-Bredenbeck, J. Süßenbach, & C. Breuer (Eds.), *Dritter Deutscher Kinder- und Jugendsportbericht. Kinder- und Jugendsport im Umbruch* (S. 24–49). Schorndorf: Hofmann.

Pack, R.-P., & Bockhorst, R. (2011). Bewegung, Spiel und Sport in Ganztagsschulen als Impulsgeber für die Entwicklung von kommunalen Bildungslandschaften. In R. Naul (Ed.), *Bewegung, Spiel und Sport in der Ganztagsschule – Bilanz & Perspektiven* (S. 164–181). Aachen: Meyer & Meyer.

Pfitzner, M., & Neuber, N. (2012). Individuelle Förderung im Sport – Didaktisch-methodische Grundlagen. In N. Neuber & M. Pfitzner (Eds.), *Individuelle Förderung im Sport – Pädagogische Grundlagen und didaktisch-methodische Konzepte* (Begabungsforschung, 14, S. 75–95). Münster: Lit.

Rauschenbach, T., Arnoldt, B., Steiner, C., & Stolz, H.-J. (2012). *Ganztagsschule als Hoffnungsträger für die Zukunft? Ein Reformprojekt auf dem Prüfstand. Expertise des Deutschen Jugendinstituts (DJI) im Auftrag der Bertelsmann Stiftung.* Gütersloh: Bertelsmann Stiftung.

Rother, P., & Stötzel, J. (2014). *Familie, soziale Herkunft und Bildungsungleichheit.* In T. Coelen & L. Stecher (Eds.), Die Ganztagsschule. Eine Einführung (S. 129–142). Weinheim: Beltz Juventa.

Salisch, M. V., Kanevski, R., Philipp, M., Schmalfeld, A., & Sacher, A. (2010). *Welche Auswirkungen hat die Ganztagsbeschulung auf die Einbindung von Jugendlichen in Peernetzwerke und Freundschaften und auf die Entwicklung sozialer und emotionaler Kompetenzen?* Unveröffentlichter Schlussbericht für das BMBF.

SpOGATA-Forschungsgruppe. (2012). *Evaluation des BeSS-Angebotes an offenen Ganztagsschulen im Primarbereich in seinen Auswirkungen auf die Angebote und Struktur von Sportvereinen, Koordinierungsstellen und die Ganztagsförderung des Landessportbundes NRW.* Essen, Münster: SpOGATA-Forschungsgruppe.

Statista (Statistikportal). (2018). *Europäische Union: Jugendarbeitslosenquoten in den Mitgliedsstaaten im April 2018.* Last access 6 June 2018: https://de.statista.com/statistik/daten/studie/74795/umfrage/jugendarbeitslosigkeit-in-europa

StEG. (2012). *Studie zur Entwicklung von Ganztagsschulen – StEG. Ganztagsschule 2012/2013. Deskriptive Befunde einer bundesweiten Befragung.* Last access 30 November 2015:http://www.projekt-steg.de/sites/default/files/Bundesbericht_Schulleiterbefragung_2012_13.pdf

Stötzel, J., & Wagener, A. L. (2014). Historische Entwicklungen und Zielsetzungen von Ganztagsschulen in Deutschland. In T. Coelen & L. Stecher (Eds.), *Die Ganztagsschule. Eine Einführung* (S. 49–64). Weinheim: Beltz Juventa.

Süßenbach, J., & Geis, S. (2013). *Evaluation von Bewegungs-, Spiel- und Sportangeboten an Offenen Ganztagsschulen in Niedersachsen.* Hannover: Landessportbund Niedersachsen.

Süßenbach, J., & Geis, S. (2014). *Kooperation Sportverein und Ganztagsschule. Ergebnisse einer Vereinsbefragung in Niedersachsen.* Hannover: Landessportbund Niedersachsen.

Süßenbach, J., & Klaus, S. (2015). Kommunale Bildungslandschaften und Sport. In W. Schmidt, N. Neuber, T. Rauschenbach, H. P. Brandl-Bredenbeck, J. Süßenbach, & C. Breuer (Eds.), *Dritter Deutscher Kinder- und Jugendsportbericht. Kinder- und Jugendsport im Umbruch* (S. 444–465). Schorndorf: Hofmann.

Suthues, B. (2014). Jugendarbeit im Sport als Gestaltungsaufgabe: Positionierungen und jugendpolitische Perspektiven. In Deutscher Olympischer Sportbund (DOSB) (Ed.), *Sport gestaltet Gesellschaft* (S. 21–33). Hamburg: Czwalina.

Thieme, L. (2012). *Ganztagsschule und Sportvereine in Rheinland-Pfalz Zentrale Ergebnisse der Bestandsaufnahme 2012 Ableitung von Handlungsempfehlungen – Executive Summary.* Last access 30 November 2015:https://www.hs-koblenz.de/fileadmin/media/hochschule/forschungsinstitute/ISS_Sportmanagement/Dateien/Executive_Summary_final.pdf

Thieme, L. (2013). Ganztag und Sportvereine. Empirische Befunde aus Rheinland-Pfalz. *Sportunterricht, 62*, 142–146.

Thieme, L. (2014). Zum Verhältnis von Ganztagsschulen und Sportvereinen in Rheinland-Pfalz. Zentrale Untersuchungsergebnisse. *Betrifft Sport, 36*(3), 10–13.

Züchner, I. (2011). Ganztagsschule und Familienleben. Auswirkungen des ganztägigen Schulbesuchs. In N. Fischer, H. G. Holtappels, E. Klieme, T. Rauschenbach, L. Stecher, & I. Züchner (Eds.), *Ganztagsschule: Entwicklung, Qualität, Wirkungen. Längsschnittbefunde der Studie zur Entwicklung von Ganztagsschulen* (S. 291–311). Weinheim: Juventa.

Züchner, I., & Rauschenbach, T. (2011). Der Sport in der Ganztagsschule. Empirische Befunde zur Kooperation von Sportvereinen und Schulen in der Ganztagsschule. In R. Naul (Ed.), *Bewegung, Spiel und Sport in der Ganztagsschule – Bilanz und Perspektiven* (S. 183–197). Aachen: Meyer & Meyer.

PART THREE:

OLD AND NEW CHALLENGES FOR THE OLYMPIC GAMES MOVEMENT

9 Sociological perspectives of the Olympic Games

March L. Krotee[23]

Modern sport and the Olympic Games in particular have developed into one of the most ubiquitous, unique and pervasive dimensions of our society. The origins of the modern Olympic Games have been vigorously chronicled by historians (Gardiner, 1925; Kieran & Daley, 1961; Lucas, 1962; Weyland, 1952). Few social scientists, however, have attempted to systematically study and offer sociological, psychosocial, or sociophilosophical interpretations of the issues, problems, trends, and future of this omnipresent sporting force (Lenk, 1976, 1979; Guttmann, 1978; Krotee, 1979a; Luschen, 1979).

From the earliest representation of the ancient Olympic Games as a symbolically sacred athletic offering to reigning deities to the most recent geopolitical representation (as portrayed by the boycott of the 1980 Moscow Games), the Olympic Games have offered an ongoing and seemingly untapped source for sociological study. During the apparent rise and demise of the Olympic Games what seems to be evident is the number of sociological factors that lend themselves to social scientific scrutiny. These include secular movements, religious factors, equal competitive opportunities, economic influences, amateurism or specialization issues, social mobility, organization, ideology, rationalization, quantification and records, and political involvement (Krotee, 1979b; Lowe, Kanin, & Strenk, 1978; Viken, Krotee, & Stone, 1979; Ueberhorst, 1976).

One such line of social scientific endeavour is examining and relating the organization and structure of the critical appendage of the Olympic Games, the International Olympic Committee, with existing sociological theory.

23 *The author wishes to acknowledge and thank the United States Olympic Committee and C. Robert Paul Jr. of Olympic House for their assistance in preparation of this manuscript.*

I wish to acknowledge and thank the United States Olympic Committee and C. Robert Paul Jr. of Olympic House for their assistance in preparation of this manuscript theory. It is the intent of this work to examine the development of the International Olympic Committee and relate its formulization to Michels's Theory of Organizational Structure in an attempt to elucidate some of the problems and issues concerning the study of this sociologic reference group.

9.1 The Rebirth of the Olympic Games

The rebirth of the Olympic Games was heavily nurtured by and committed to the lofty ideals of equality, peace, and good will of their creator, Baron Pierre de Coubertin. Coubertin stated that if the German Reich had excavated the ruins of ancient Olympia through an 1875 pact with the Greek government, why then should France not restore the old splendour? From here it was not too far to a less splendid but fertile and practical idea to renew the Games, especially because it seemed that international sport should play a new role in the world (Coubertin, 1896, 1908). As Coubertin idealistically, humanistically, nationalistically, and optimistically envisaged, youth, education, and athletics were intricately bound; international rivalry in sports would promote international amity in broader fields, and all humankind would be benefited.

Thus, the inherent struggle over the Olympic ideal reappeared at the Grand Hall of the Sorbonne in Paris at the International Athletic Congress on June 16, 1894. Also resurfacing was the activating force that resulted in the reestablishment of the Olympic Games. The congress was to have served as a forum to discuss the growing conflicts within modern sport including rules and regulations, and the amateurism/professionalism dilemma dominated its ten-item agenda. What emerged, however, was the decision to establish a 14-member International Olympic Committee and to revive the ancient Olympic Games and conduct them in Athens in 1896. Coubertin stated as follows:

> *The re-establishment of the Olympic Games based upon the conditions conforming to the needs of modern life would serve to bring together every four years*

representatives of all nations, and it is permissible to suppose that these peaceful and courteous contests would supply the best of internationalism.

Cited in Leonard & Affleck, 1957

The response to Coubertin's appeal was described by Professor William Milligan Sloane of Princeton University, one of the original members of the hand-picked International Olympic Committee. The Belgians were hostile, the Britons were lukewarm, the uninvited Germans represented by Baron von Reiffenstein took no notice whatsoever, and the French vowed not to send a team if the Germans were invited. Even the Greek government was financially sceptical of becoming the recipients of the dream that had been born and had flourished in Olympia.

Nevertheless, on the 75th anniversary of Greece's declaration of independence from Turkey, on a rainy Wednesday morning on April 6, 1896, in the lavishly remodelled stadium of Herodis, King George I of Greece formally opened the first modern Olympic Games. The ancient Greek Olympic Games that had been halted by the Roman emperor Theodosius I had been revived after a lapse of 14 centuries by an idealistic, patriotic, and nationalistic Frenchman who believed in sport for all (except women) and who seemed to bury his egalitarian passions in a quadrennial festival for the elite athlete. Thus, the modern Olympic Games emerged as the ideal mixture of physical prowess, moral attitude, religion, beauty, and peace for the betterment of humankind; a nineteenth-century dream seemingly already in conflict with the emerging transnational and dilemmatic twentieth century.

9.2 A sociological view of the International Olympic Committee

An analysis of the International Olympic Committee may also serve as an introspective testament to the patriotic intensity and tenacious psychosocial profile of its creator and mentor from the year of the first modern Olympic Games until 1925 – Baron Pierre de

Coubertin. Although he was always the champion of individual freedom, human rights, democracy, moral humility, and integrity, Coubertin, on June 13, 1894, at the International Athletic Congress in Paris, personally selected 14 men to serve as the foundation for the reconstruction of the modern Olympic Movement as he believed it should be constructed. Coubertin preached that

> *the foundation of real human morality lies in mutual respect – and to respect one another it is necessary to know one another . . . Let us export oarsmen, runners, fencers; there is the free trade of the future – and on the day when it shall take place among the customs of Europe the cause of peace will have received a new and powerful support.* United States Olympic Committee, 1979b

Thus, Olympism sought to contribute all that sport could to the social, educational, moral, and aesthetic integrity of the quality of life of the individual.

The original members of the International Olympic Committee charged to carry out Coubertin's will included Demetrios Vikelas (Greece), Ernest Callot (France), General Alexander D. von Butowski (Russia), Colonel Gustav Balck (Sweden), Professor William Milligan Sloane (United States and Princeton University), Dr. Jiri Guth-Jarkowski (Bohemia, at this time part of the Danube-Monarchy), Ference Kemeny (Hungary), Lord Amphtill and George Herbert (Great Britain), Dr. J. B. Zubiaur (Argentina), Leonard C. Cuff (New Zealand), Count Lucchesi Palli and Earl Andria Garafa (Italy), and Count Maxime de Bousies (Belgium). Even though the official interlocking rings of the Olympic flag, flown initially in Antwerp in 1920, symbolize the union of the five continents, the original International Olympic Committee membership was dominated by the selection of representatives of the European sporting nations and, for the most part, special acquaintances of the baron. This slight flaw in democratic procedure made little difference in the internal mechanism of the committee because Coubertin reigned over the International Olympic Committee, was sole director of the Games' form and character, developed the Olympic Charter, Olympic Protocol, and the athlete's oath as well as the format for the opening and closing ceremonies. He personally assumed all administrative and financial duties. Even today, few major alterations from his legacy have emerged from this self-elected, self-anointed, and self-perpetuated entity where members are not delegates from countries but delegates to countries.

For the social scientist, the organization and structure of the International Olympic Committee is worth sociological inspection. The committee seems to be a formal, rationally organized bureaucratic social structure as described by Weber. This includes possessing well-defined goals, interests and purposes, clear patterns of activity, and maintaining acceptable modes of regulations and control that functionally relate to the purpose of the organization (Cassenelli, 1953; Melser, 1971; Merton, 1968; Parsons, 1951). A closer inspection of the International Olympic Committee, however, reveals that the committee patterns itself after Michels's (trans. 1959) Theory of Organizational Structure. Michels (1876–1936) was a German-Italian sociologist who studied cooperative societies, social movements, and various internal structures of Western European political groups. From his observations, he found that these groups, despite their commitment to the ideal of equality, failed to maintain these ideals in their own structures and instead developed rigid and rather permanent oligarchies. He thus formulated the 'iron law of oligarchy'. Michels believed that certain fundamental sociological laws prohibited the attainment of social equality no matter what economy or political system prevailed. He believed that the power of the organization was in its political makeup and influence of its members.

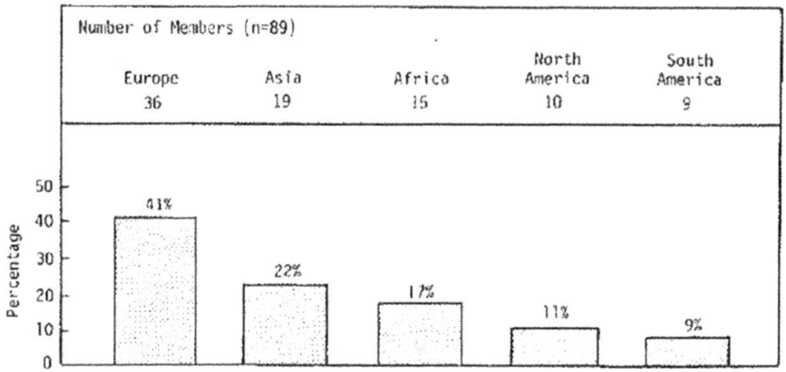

Figure 9.1: The membership of the International Olympic Committee by continent, Krotee (Note 1).

Thus, in examining the organization, structure, and membership of the International Olympic Committee, one needs only to examine its mailing address – Chateau de Vidy,

1007 Lausanne, Switzerland, an eighteenth-century mansion overlooking Lake Geneva – to detect a slight air of elitism. Further study of the committee membership (see appendix A) reveals such personages as the Marquess of Exeter, Cheik Gabriel Gemayel, Prince Regnant Francois Joseph II of Liechtenstein, Prince Gholam Reza Pahlevi, and the Rajah Bahalindra Singh as well as assorted dukes, counts, barons, and generals. This membership does little but confirm the spirit of democratic elitism. Like the initial committee membership selected by Coubertin, today's International Olympic Committee, as well as its past presidents, seems to be dominated by the elite upper class and, for the most part, European patrimony (see fig. 9.1, table 9.1).

The membership of the committee is self-recruited and is restricted to those who speak French or English and are citizens of and reside in a country that possesses a recognized National Olympic Committee. Technically, only one member from any country should be on the committees, including countries where the Games have been hosted, except for the largest and most active countries in the Olympic Movement; The members are representatives of the International Olympic Committee in their countries and not their countries' delegates; they may not accept from governments any instructions that would interfere with the independence of their vote. Membership ceases if a member resigns, becomes inactive, reaches the age of 72 if elected after 1965, changes nationality, or is deemed unworthy by the committee.

Table 9.1: The presidents of the International Olympic Committee, 1894–1980

First	Demetrias Vikelas	Greece	1894–1896
Second	Baron Pierre de Coubertin	France	1896–1925
Third	Count Henry de Baillet-Latour	Belgium	1925–1942
Acting President	J. Sigfrid Edström	Sweden	1942–1946
Fourth	J. Sigfrid Edström	Sweden	1946–1952
Fifth	Avery Brundage	United States	1952–1972
Sixth	Lord Michael Killanin	Ireland	1972–1980
Seventh	Juan Antonio Samaranch	Spain	1980–present

The International Olympic Committee consists of a president, who is elected by secret ballot for an eight-year term; three vice presidents, who serve four-year terms; and an executive board of the four officers and five additional members (see appendix B). This body insures the rules are observed, is directly responsible for the administration and management of finances, decides all matters of doubt or dispute concerning the Olympic Games and the Olympic Movement, and interprets the rules and applies penalties to organizations and individuals under its jurisdiction. Another segment of the International Olympic Committee is the Olympic Congress (or Tripartite). The Olympic Congress consists of the International Committee membership (see appendix A), delegates of the International Federations (see appendix C), and the recognized National Olympic Committees (see appendix D).

The juridical objectives and powers of the International Olympic Committee remain today much as they have been in the past. The International Olympic Committee as described by the Olympic Charter that was adopted in Montevideo, Uruguay, April, 1979, is a nonprofit body, corporate by international law, and has juridical status and perpetual succession. Its aims are to (a) encourage the organization and development of sport and sport competition; (b) inspire and lead sport within the Olympic ideal, thereby promoting and strengthening friendship between the sportsmen of all countries; (c) ensure the regular celebration of the Games; and (d) make the Olympic Games ever more worthy of their glorious history and of the high ideal that inspired their revival (Comite Internationale Olympique, 1979).

The International Olympic Committee is financed by the cities entrusted with organizing the Games as well as by monies accrued from the Games. Its jurisdiction includes such areas as entry, eligibility, program, participation of women, events, equipment, art exhibits, traveling expenses, housing, technical aspects, prizes, media, choice of site, time and duration, and music – in short, all aspects of the Games. In the final analysis, the International Olympic Committee, currently consisting of 89 members (see appendix A), has supreme authority on all questions concerning the Olympic Games and the Olympic Movement. The Amateur Sports Act of 1978 (Federal Law 95-606) enacted by President Carter on November 8, 1978, further designates the United States Olympic Committee as the central coordinating agency in the United States for all sports in Olympic programs (United States Olympic Committee, 1979a).

9.3 Michels's theory applied

The structure, organization, function, and moralistic doctrine of the International Olympic Committee provide the democratic and idealistic egalitarian conceptual framework to which Michels's theory may be applied. Michels's theory is that oligarchy or a form of organization in which the power is vested in a few arises in even the most idealistic of all organizations. He believed that ideals and ideology serve only as rationalizations to preserve a position of power within a social organization. Michels studied the interaction between the nature of the social organization and the individuals involved, particularly the size, complexity, or organizational function and the coordination of group activities. Also, of significance were the psychosocial tendencies of the organization's leaders and followers, which he felt were critical to the formation of an oligarchic structure. The interaction of the above factors is expressed in Figure 9.2.

As shown previously, the fundamental range of Michels's theory as applied to an examination of the International Olympic Committee seems to result in a semblance of 'goodness of fit'. The International Olympic Committee and its associated affiliations comprise a large number of groups and individuals representing omnifarious forms of techno-sport (see appendices C and D). The complexity of the coordinated functioning and interrelationships of these multidimensional organizations with the International Olympic Committee represents a Herculean charge (Nafziger, 1971). These technical and administrative features of the structure of the International Olympic Committee serve to operate against democratic determination, and thus, reflect Michels's model of causal relationships contributing to the development of an oligarchical structure.

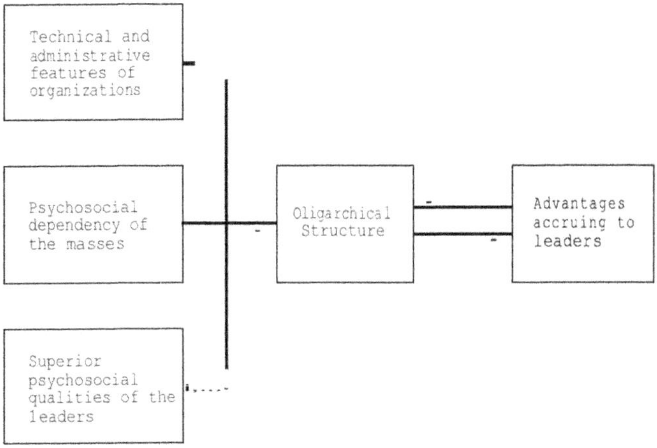

Figure 9.2: Causal relationships among variables contributing to the development of an oligarchical structure. (Adapted from Smelser, 1971.)

The psychosocial dependency of the public on the continuation and conduct of the Olympic Games has been evident since 1894, when Baron de Coubertin and Crown Prince Constantine financially revived the seemingly fiscally doomed inaugural Games destined for Athens.

The Games have since mushroomed into a multimillion-dollar business venture administered by the International Olympic Committee, which delegates various aspects of its role through a labyrinth of appointed committee substructures. Thus, Michels purported that the mechanical and technical possibility of direct governing of the Olympic Games by the public is impractical. This is due to a myriad of reasons, such as the voluntary and transnational nature of the organization and the need for constant communication and coordination. These factors ne gate the equal involvement of all members even within the International Olympic Committee itself. As a result, the technical specialization of what Michels refers to as expert leadership is requisite. Such are the origins of centralized power and the formulization of oligarchy, where all related variables (as in fig. 9.2) seem to slide in one focal direction. It reflects an organization that gives birth to the dominion of the elected over the electors, of the mandatories over the mandators, and of the delegates over the delegators. It reflects an organization where countertendency does not thrive and a dependency of the masses is created due to the enormous amount of effort, time, and monies required to ensure a successful Olympic outcome.

To ensure this success and perpetuate itself, an organization with the magnitude of the International Olympic Committee needs the direction and guidance of superior leadership. The psychosocial attributes of superior and domineering leadership that has been reflected throughout the history of the International Olympic Committee is also an integral component of Michels's causal conceptual framework. Michels mentioned leaders with strength of conviction, pride, force of ideas, moral correctness, and extreme dedication, and certainly these attributes have been more than characterized by the past and present leadership of the International Olympic Committee (see table 9.1, appendices A and B). These attributes seem to expand as do the self-adulation and the notion of indispensability of the leaders as their years in office unobtrusively accumulate. The leaders of the International Olympic Committee remain securely placed, and psychologically gratified; thus, they possess information, cultural accessories, and wealth – all of which fit snugly into Michels's conceptual framework. Michels's theory also suggests that a minority of leadership gives orders to the majority, that the minority of leaders are the primary source of political action, that individuals in these positions of power may pursue their own interests at the exploitation of others, and that those leaders tend to consolidate their positions over long periods of time.

Michels's theory also allows for further and future attempts by the committee to maintain credibility, to ensure that the organization still identifies itself with democracy and the Olympic ideal, and to represent the good of all. The International Olympic Committee's establishment of a network of external commissions and committees to investigate itself and to make recommendations to itself indicates such attempts.

One recently formed committee is concerned with discrimination against female participation in the Games, despite the Olympic ideal reflected in Rule 3 of the Olympic Rules that purports to develop both boys and girls, men and women through competitive sport and physical training. Table 9.2 illustrates the historic flow of male and female Olympic participation patterns reflecting the continued discrimination against female athletes by the International Olympic Committee and its complex web of substructures. The International Olympic Committee has also established committees to consider enlarging and equalizing the committee, to investigate discriminatory practices against developing and socialist countries, and to ensure more input and establish closer relations with international federations and the National Olympic Committees.

Table 9.2: Historical chart of the modern Olympic Games (Courtesy of the United States Olympic Committee)

	Olympiad Site – dates of competition	No. of nations	No. of events	Men's sports	Men athletes	Women's sport	Women athletes	Total no. of athletes
I	Athens, Greece – April 6–15, 1896	10	42	10	285	0	0	285
II	Paris, France – May 20–Sept. 20, 1900	20	60	12	1,060	1	6	1,066
III	St. Louis, MO – July 1–Oct. 15, 1904	10	67	10	496	0	0	496
IV	London, England – April 27–Oct. 27, 1908	22	104	17	2,023	3	36	2,059
V	Stockholm, Sweden – May 5–July 19, 1912	28	106	15	2,484	3	57	2,547
VI	Berlin, Germany – not celebrated – World War I							
VII	Antwerp, Belgium – July 22–Sept. 5, 1920	29	154	19	2,543	3	63	2,606
VIII	Paris, France – May 4–July 27, 1924	44	137	18	2,956	3	136	3,092
IX	Amsterdam, Netherlands– May 17–Aug. 12, 1928	46	120	16	2,725	4	290	2,015

(Continued)

(Continued)

Table 9.2: Historical chart of the modern Olympic Games (Courtesy of the United States Olympic Committee)

	Olympiad Site – dates of competition	No. of nations	No. of events	Men's sports	Men athletes	Women's sport	Women athletes	Total no. of athletes
X	Los Angeles, CA – July 30–Aug. 14, 1932	37	124	17	1,281	3	127	1,408
XI	Berlin, Germany – Aug. 1–16, 1936	49	142	19	3,741	4	328	4,069
XII	Tokyo and Helsinki – not celebrated – World War II							
XIII	London, England – not celebrated – World War II							
XIV	London, England – July 29–Aug. 14, 1948	59	138	19	4,304	5	385	4,689
XV	Helsinki, Finland – July 18–Aug. 3, 1952	69	149	19	4,925	5	518	4,925
XVI	Melbourne, Australia* – Nov. 22–Dec. 8, 1956	67	145	19	2,813	5	371	3,184
	*Stockholm, Sweden – June 10–17, 1956	29	3	1	132	1	13	145
XVII	Rome, Italy – Aug. 26–Sept. 11, 1960	84	150	18	4,859	5	537	5,396
XVIII	Tokyo, Japan – Oct. 10–24, 1964	94	162	20	4,826	8	732	5,558
XIX	Mexico City, Mexico – Oct. 12–27, 1968	109	172	19	5,238	8	844	6,082

(Continued)

Table 9.2: Historical chart of the modern Olympic Games (Courtesy of the United States Olympic Committee)								
	Olympiad Site – dates of competition	No. of nations	No. of events	Men's sports	Men athletes	Women's sport	Women athletes	Total no. of athletes
XX	Munich, Germany – Aug. 26–Sept. 10, 1972	108	196	21	7,019	9	1,125	8,144
XXI	Montreal, Canada – July 17–Aug. 1, 1976	88	199	21	4,915	10	1,274	6,189

* Equestrian sports, only, condiucted in Stockholm because of immigration laws for horses in Australia

Table 9.2 Historical chart of the modern Olympic Games (Courtesy of the United States Olympic Committee)								
	Olympiad Site – dates of competition (Winter Games)**	No. of nations	No. of events	Men's sports	Men athletes	Women's sport	Women athletes	Total athletes
I	Chamonix, France – Jan. 25–Feb. 4, 1924	16	16	5	280	1	13	293
II	St. Moritz, Switzerland – Feb. 11–19, 1928	25	15	5	464	1	27	491
III	Lake Placid, NY – Feb. 4–13, 1932	17	19	5	277	1	30	301
IV	Garmisch-Partenkirchen, Germany – Feb. 6–16, 1936	28	21	5	680	2	76	756
V	St. Moritz, Switz. – Jan. 20–Feb. 8, 1948	28	24	6	636	2	77	713

(*Continued*)

(Continued)

Table 9.2 Historical chart of the modern Olympic Games (Courtesy of the United States Olympic Committee)

	Olympiad Site – dates of competition (Winter Games)**	No. of nations	No. of events	Men's sports	Men athletes	Women's sport	Women athletes	Total athletes
VI	Oslo, Norway – Feb. 14–25, 1952	30	23	5	624	2	108	732
VII	Cortina, Italy – Jan. 26–Feb. 5, 1956	32	24	5	778	2	146	924
VIII	Squaw Valley, CA – Feb. 18–28, 1960	30	27	5	534	3	159	693
IX	Innsbruck, Austria – Jan. 20–Feb. 9, 1964	36	28	7	1,107	4	225	1,332
X	Grenoble, France – Feb. 6–18, 1968	37	35	7	1,129	4	230	1,359
XI	Sapporo, Japan – Feb. 3–13, 1972	35	35	7	927	4	219	1,146
XII	Innsbruck, Austria – Feb. 4–15, 1976	37	37	7	1,013	4	218	1,261
XIII	Lake Placid, NY – Feb. 13–24, 1980		38	7		4		

** Previously some winter events were conducted as part of the summer Olympic Games.

Recent steps such as appointing Committee delegates who are younger, less socially conspicuous and who possess sporting ties, and encouraging referendums from various substructures (see appendices C and D) also fit into Michels's schema.

9.4 The future of the Olympic Games

As events concerning the boycotted 1980 Olympic Games continue to unfold and more problems and questions confront the International Olympic Committee, only time will tell if the present oligarchic power structure of the International Olympic Committee, which may have been needed to reinstate the Olympic Games in the earlier years of the Olympic Movement, can withstand the pressures of true democratic spirit. It remains to be seen if, as Michels's theory suggests, an organization that compromises and enters into promiscuous relationships with other elements in society sacrifices its political virginity, which may have disastrous and enduring consequences. In this instance, the consequences are the life of the Olympic Games and the Olympic Movement itself.

Unless the Olympic forum is radically shaken, however, it appears that the International Olympic Committee will not lose its grip on the Olympic olive branch or, indeed, the Olympic forum will not be altered to allow for harmonious direction of wills toward identical objectives and practical aims. When such change is lacking, an organization no longer contributes to society and eventually becomes ineffective and dissolves. It appears that the Moscow Games may be such a turning point in history. Ultimately, however, the effects of the 1980 boycott remain to be determined.

Reference Note

Krotee, M. L. *Sport sociology: Issues and problems of the Olympic Games.* Paper presented at the American Alliance of Health, Physical Education, Recreation and Dance Conference. Detroit, MI, April 10, 1980.

References

Beck, R. H. (1974). The Greek tradition and today's physical education. In C. A. Bucher (Ed.), *Dimensions of physical education.* St. Louis, MO: C.V. Mosby.

Cassenelli, C. W. (1953). The law of oligarchy. *The American Political Science Review, 47,* 773–784.

Comite International Olympique. *Olympic Charter 1979.* Lausanne, Switzerland: Comite International Olympique.

Coubertin, P. (1896). *The Olympic Games of 1896.* New York.

Coubertin, P. (1908). Une campagne de vingt-et-un ans 1887–1908. Paris: Librarie de L'Education Physique.

Gardiner, E. N. (1925). *Olympia: Its history and remains.* Oxford: Clarendon Press.

Guttmann, A. (1978). *From ritual to record.* New York: Columbia University Press.

Kieran, J., & Daley, A. (1961). *The story of the Olympics.* Philadelphia: J.B. Lippincott.

Krotee, M. L. (Ed.). (1979a). *The dimensions of sport sociology.* West Point, NY: Leisure Press.

Krotee, M. L. (1979b). The rise and demise of sport: A reflection of Uruguayan society. *The Annals of the American Academy of Political and Social Science, 445,* 151–154.

Lenk, H. (1976). Toward a social philosophy of the Olympics: Values, aims and reality of the modern Olympic Movement. In P. J. Graham & H. Ueberhorst (Eds.), *The modern Olympics.* West Point, NY: Leisure Press.

Lenk, H. (1979). *Social philosophy of athletics.* Champaign, IL: Stipes.

Leonard, F. E., & Affleck, G. B. A (1957). *Guide to the history of physical education.* Philadelphia: Lea & Febiger.

Lowe, B., Kanin, D. B., & Strenk, A. (1978). *Sport and international relations.* Champaign, IL: Stipes.

Lucas, J. (1962). *Baron Pierre de Coubertin and the formative years of the modern international Olympic Movement.* Unpublished doctoral dissertation, University of Maryland.

Luschen, G. (1979). The organization and policies of National Olympic Com mittees-A pilot project. In M. L. Krotee (Ed.), *The dimensions of sport sociology.* West Point, NY: Leisure Press.

Merton, R. K. (1968) *Social theory and social structure.* New York: Free Press.

Michels, R. (1911). *Political parties* (E. Paul & C. Paul, trans.). New York: Dover (trans. 1959).

Nafziger, J. A. R. (1971). The regulation of transnational sports competition: Down from Mount Olympus. *Vanderbilt Journal of Transnational Law, 5*, 180–212.

Parsons, T (1951). *The social system.* New York: Free Press.

Smelser, N. J. (1971). *Sociological theory: A contemporary view.* New York: General Learning Press.

Ueberhorst, H. (1976). Return to Olympia and the rebirth of the games. In P. J. Graham & H. Ueberhorst (Eds.), *The modern Olympics.* West Point, NY: Leisure Press.

United States Olympic Committee. (1979a). *Constitution and by-laws.* Colorado Springs, CO: United States Olympic Committee.

United States Olympic Committee. (1979b). *The Olympic Games.* Colorado Springs, CO: United States Olympic Committee.

Viken, J. P., Krotee, M. L., & Stone, G. P. (1979) Popular culture and the erosion of class distinctions in the mass sport ot drag racing. In M. L. Krotee (Ed.), *The dimensions of sport sociology.* West Point, NY: Leisure Press.

Weyland, A. M. (1952). *The Olympic pageant.* New York: MacMillan.

10 Olympism and Olympic education: Back to the roots

Michael Krüger

Roland Naul is one of those physical educators who never get tired of charming the relevance of Olympism for general education, despite the scandals coming along with the Olympics: doping first, followed by deception and corruption, and in a wider context the destruction of Olympic values through commerce without restraint, through media and politics, who do not respect the integrity of sport, and not least through Olympic officials themselves, who are neither aware of the Olympic ethos and tradition nor of their responsibility toward humans and humanity, but only looking out for themselves.

These are major reasons why most people associate Olympic sports and Olympic Games less with moral and pedagogical ideals, and more with show business, doping, commerce, and corruption, in other words, less with fair play and more with foul play. Many people bitterly resent and reject the decline of sports in the context of commerce, corruption, and show business. They want and indeed demand integrity and fairness in sport. Roland is one of those who rather believe in the chances of Olympism and work for overcoming the destruction of a big hope of humanity since the twentieth century.

Bearing in mind the history of modern Olympic Games and the apparently growing gap between ideals of its fundamental principles, and the reality, I will firstly attempt to explain in a historical perspective what Olympism really means, and how its understanding has changed during the last century. The process of renewing Olympism can only succeed by remembering and reassuring its roots in history, ethics, and education.

Secondly, I will discuss three perspectives of Olympic education, as crucial elements of modern education as a whole.

10.1 What exactly is Olympism?

The first Olympic Games of modern times took place in Athens, Greece, in 1896, nearly 120 years ago. Pierre de Coubertin, a French nobleman, is usually looked upon as the inventor of this new, modern tradition with its respect for antiquity. However, he was not alone. At that time, many educated people of various European nations like Great Britain, France, Greece, and last not least Germany, were full of respect and enthusiasm for classic Greek antiquity of more than 2,000 years ago. For them, this period seemed to be the peak of both humanity and humanism, and of cultural development. In modern times, on the eve of the twentieth century, the looking back at Greek antiquity was expected to provide the right orientation for the future. Therefore, in Europe, this period is referred to as new humanism.

Pierre de Coubertin had argued that Olympic Games in modern times should be no more than a revival of the ancient Olympic Games.[24] In fact, these events in antiquity were no games at all but religious ceremonies including athletic contests. In 1894, when Coubertin had invited some influential gentlemen from Europe and the United States of America for a congress on amateur sport, he placed as last top of the agenda: 'On the opportunity to reinstall Olympic Games'. He used the terms *Olympic*, *athletic*, and *gymnastic*, in order to ennoble modern sports with classic culture. He also distinguished between such modern sports as athletics, gymnastics, swimming, rowing, and horse riding, which seemed to parallel classic disciplines in antiquity and others like ball games, which were not part of the classic tradition of athletics and agonistics. As in antiquity, the single male athlete contesting in athletic competitions should be the star or idol of modern Olympics.

In fact, the congress in Paris in 1894 marked the birth of modern Olympic Games, of modern Olympic sport organizations and institutions and, finally, of a new Olympic ideal of modern sport. The International Olympic Committee (IOC) was founded, and the first

24 *Coubertin reported several times of the Sorbonne congress, that is, Pierre de Coubertin,* Einundzwanzig Jahre Sportkampagne (1887–1908): (Une campagne de vingt-et-un-ans), *with the assistance of B. Wirkus (Ratingen, Kastellaun, Düsseldorf: A. Henn Verl., 1974), pp. 74–81, and Pierre de Coubertin,* Olympische Erinnerungen, *Reprint der 2. Aufl. 1959 (Wiesbaden: Limpert, 1996), pp. 13–25.*

president of this new international Olympic Sports organization was Dimitrios Vikelas, who would organize the first Games in Athens. In fact, Pierre de Coubertin was both *the spiritus rector* and general manager of the modern international Olympics. At first and in general, the main purpose of the IOC was to organize Olympic Games in the rhythm of four years – like the Olympians in antiquity – but by contrast to antiquity not always at the same place, but in different cities of the world. Consequently, president of the IOC ought to be the organizer of the next Games. Meanwhile, the IOC developed into the very powerful and wealthy 'world government of sport', as the political scientists Rittberger & Boekle called it.[25]

Incidentally, at present and for the first time in Olympic history, a German sportsman, Thomas Bach, is president of the IOC. Despite the complicated relationship of the Germans to the Olympic Movement during history, this fact can be regarded as a clear indication that Germany has finally arrived in Olympia. The IOC, as head of the international Olympic family, has succeeded in uniting 'all games' and 'all nations' (or 'all sports for all people') in the spirit of Olympic sport, as Coubertin once proclaimed this major aim of the Olympic Movement.[26] However, the spirit of sport apparently changed over the last century. Whereas the Olympic vision of 'all games, all nations' seems to have been achieved, the original Olympic ideals like amateurism have truly vanished.

Coubertin himself not only wished to organize a new international sports event, he wanted more. He intended to create a new 'religion' of modern Olympic sport, based on the ideals of antiquity. By working toward this vision, he was well aware of the fact that modern Olympic Games had to be different from those of antiquity, 2,000 years ago. They had to confirm to the conditions and needs of modern times. Coubertin wanted to ground modern Olympic Games in an idea beyond athletic or sporting practices. That is

---

25 *Volker Rittberger and Henning Boekle, "Das Internationale Olympische Komitee – eine Weltregierung des Sports?," in* Olympischer Sport: Rückblick und Perspektiven, *ed. Ommo Grupe (Schorndorf: Hofmann, 1997).*

26 *Pierre de Coubertin propagated this branding on various occasions, for example in the Olympic letter XXI, May 17, 1919, p. 1, in Norbert Müller, ed.,* Pierre de Coubertin, 1863–1937: Olympism *(Lausanne: IOC, 2000); Selected Writings, p. 551.*

what he called Olympism, the Olympic idea or Olympic philosophy, and sometimes even *religio athletae*, which means Olympic religion.

It is worth noting that this difference between practical Olympic sport in the stadium and sporting halls on the one hand, and theoretical reflections or concepts in this context on the other hand, was unknown in antiquity. There was no need to legitimize athletic practices in antiquity, although the Olympians had been celebrated for about 1,000 years. Athletics and agonistics were a matter of course in the context of ritual and cultic actions during such events. In antiquity, they were part of the ceremonies to praise the Hellenic gods by making various sacrifices.[27] Authors and philosophers of the antiquity commentated on and criticized the behaviour of single athletes or specific elements of the events, but there is no specific and coherent philosophy of the antique Olympians or Panhellenic agons, compared to the philosophy of the modern Olympic Games.

Coubertin was the first to define the meaning of Olympism in modern times, or perhaps rather what he himself wanted to be the substance and the deeper sense of his new sporting religion. Until today, various thinkers, educators, and philosophers all over the world attempt to define, interpret, and understand modern Olympism. Not least, the IOC, in its role as heir to Coubertin and owner of the Olympic Games, its labels and symbols, again and again tries to adopt and modernize the basic Olympic message.

In 1896, just before the first Games in Athens, Coubertin published an article in the second Edition of the official journal *Revue Olympique* defining Olympism, or, as he wrote, the 'essence of our critical endeavor'. He explains:

> *Our idea to revive such an institution which was forgotten for many centuries, is the following [. . .]. The relevance of athletics is growing year by year. Its role in modern times seems to be as relevant and sustainable as in ancient times. In addition, athletics acquired modern characteristics. It is international and democratic, which means adopted to the ideas and needs of the present. However, today*

27 *Emanuel Hübner,* Das Olympische Dorf von 1936: Planung, Bau und Nutzungsgeschichte, *Aufl. 2015 (Paderborn: Schöningh Paderborn, 2015).*

as well as in the past, its effects may be healthy or harmful, depending on the use of athletics and the way we perform sports and athletics. Athletics can produce noble as well as base excitement and emotions; it may develop a sense of honor and altruism, as well as egoism and greed (mammonism). Sports and athletics can be fair or foul, both functionalized to secure peace and to prepare for war.

Coubertin, Olympische Erinnerungen, p. 27

Coubertin was aware of the fact that sport and athletics are not morally useful and worthy in their own right, but only through the appropriate educational interventions and structures. Modern sport ought to be used as a means of education. Therefore, Olympic sport contributes to a better and more human education in the interests of peace and harmony. Of course, sport and athletics in modern times ought to be exercised fairly, respectfully, in friendly partnership with others and in a harmony of body, mind, and will.

Coubertin respected these characteristics that are typical of modern, 'democratic societies', as he added the following: 'Noble emotions, culturing a spirit of honor and altruism, fair play, male strength and peace are the first needs of modern democratic societies, irrespective of whether they are republican or monarchic.'

What did Coubertin think about 'male power' or 'male strength'? He was convinced that the single fighting male athlete should be the idol of modern Olympic sport. As in antiquity, women should not participate in Olympic Games, but praise the athletic male heroes and winners of the competitions. For him, young strong male athletes symbolized progress and the future for both the individual and society as a whole, the nations, and finally the progress of the human race.

In contrast to the anachronistic understanding of Coubertin, meanwhile the participation of women in the Games is a matter of course. This fact may be one of the most important changes in modern Olympic history. Simultaneously, the growing relevance of women in the Olympic Movement may be one of the most serious errors of Coubertin, next to that of amateurism.

In 1935, one year before the XIth Olympic Games of Berlin 1936, Coubertin tried to explain again, and it was to be his last opportunity to do this, what Olympism means. At that time, the Olympic Games were established in the world, well known and popular in many modern, civilized nations. The German organizers of the Games asked Coubertin to give a speech on world radio for an international audience. This speech on 'the philosophical basics of modern Olympism' was given in French, but translated into various languages.

Coubertin's radio speech in 1935 can be regarded as a kind of integration of all the ideas, experiences, and reflections about Olympic sports during his long life for the cause of Olympic sport and athletics. He distinguished between four issues, in order to explain Olympism:[28]

1. "The first and essential criterion of both the old and modern Olympism is: to be a religion."

   What did Coubertin mean by this notion of Olympism as a religion? In fact, Coubertin was wrong in his assertion that there had been a religion of Olympism in antiquity. In reality, athletics and agonistics were part of religious cults, but athletics for its own sake was unknown in antiquity. However, which sense does religio athletae make with respect to modern Olympics and in the present?

   There were various ways of understanding or interpreting this strange and, in some respect, blasphemous notion. I think that Coubertin was considering the religious feelings of athletes in antiquity by praising their gods through *agons,* and comparing these with motivations and emotions of athletes at modern Olympics. They should, he believed, be aware that their actions during training and competition are by no means 'just for fun', as they are ambassadors of the message of modern Olympism, prophets or priests of the *religio athletae,* communicating with their body and performance. This message includes a universal idea of education for 'democracy and internationalism', as Coubertin never ceased to point out. Top athletes

28 *The following quotes are translated from the original transcript of the French speech Pierre de Coubertin and Organisationskomitee für die XI. Olympiade Berlin 1936 e.V.,* Pax Olympica: Weltsendung des Reichssenders Berlin am Sonntag, dem 4. August 1935 mittags, *with the assistance of Pierre d. Coubertin, and Carl Diem (Berlin-Charlottenburg: Organisationskomitee für die XI. Olympiade Berlin 1936 e.V., [1935]), [Vorolympische Kampagne in drei Sprachen].*

at Olympics are idols for a new generation and encourage them to create a better world of progress and humanity. They are idols for all, to strive for better achievement, discipline, and self-control, in order to foster a peaceful world characterized by democracy and internationalism.

2. "The second criterion of Olympism refers to nobility and selection, but a kind of nobility which includes complete equality from the beginning. This Olympic nobility is based only on bodily predominance of the single athlete, his progressive bodily variety, and to a certain degree on his will to train and exercise."

This second Olympic principle includes a clear commitment to the principle of achievement and performance. Striving for excellence is essential for Olympic sports. 'Citius, altius, forties,' faster, higher, stronger, is the official motto of the Olympic Movement. This motto is related to one of the oldest poems of humankind, the *Iliad* of the antique writer Homer. 'Striving ahead and to be better than the others' ('Vorwärts streben und besser sein als die anderen') was the fundamental principle of life for the old Greeks, and, as some historians assert, the miracle of their high and never-again-reached level of cultural development.[29]

In this sense, Olympic athletes are idols for all, encouraging them to do their best, to achieve their human potential, in order to contribute to the progress of humankind, based on equality and personal achievement. Competitions are necessary in order to select the best, the Olympic nobility. Competitions are played in the spirit of fair play, a principle not limited to athletes and including fans as well. Their applause should only be a response to pure achievement and not for other reasons like nationality, sympathy, religion, birth, race, or any other factor.

3. 'The idea of peace is an essential element of Olympism as well, similar to the idea of eccecheiria in antiquity. This idea is connected to that of rhythm.'

---

29 *For example, Jakob Burckhardt,* Griechische Kulturgeschichte, *2. Aufl., 13.-15. Tsd, Dtv 6078,4 (München: Dt. Taschenbuch-Verl, 1982 (Original 1898–1902)) and the philosopher Friedrich Nietzsche, especially in his essay Friedrich Wilhelm Nietzsche and Christa Davis Acampora,* Homer's contest, *Nietzscheana / North American Nietzsche Society 5 (Urbana, IL: North American Nietzsche Society, 1996).*

The Olympic Games were to be celebrated every four years, according to the antique tradition. The Games symbolize the 'four-year festival of human spring-time', as Coubertin said. Olympic Games are festivals for the young generation, bearing the hope of a better and more peaceful world in the future. Young athletes build bridges between individuals, nations, and generations, between the past and the future. 'How could we give honor to them better' (to the young athletes, MK), asked Coubertin rhetorically, 'than through the common demand for peace, at least during the Olympic Games!'

The German sociologist Helmut Schelsky translated the Greek *eccecheiria* as 'temporary peace' *(Friede auf Zeit).*[30] Coubertin himself was realistic enough to see that peace in general is not compatible with human nature. He was not a pacifist, but he believed that Olympic education should contribute to better self-control and, in the long run, to peace, based on strength. Sporting contests are examples of the educational opportunity to learn how to deal with conflicts in a fair and peaceful way. 'A man is strong indeed,' said Coubertin, 'when he succeeds in both overcoming himself and stopping imperiousness and greed for the property of others.'

4. 'Finally, the last element of Olympisms is beauty through unity of art and mind at the Olympic Games.'

The Olympic Games ought to be a total work of art composed of bodily sports and competitions at the centre, and supplemented by fine arts, music, literature, science, philosophy, and history. 'To celebrate Olympic Games includes relating to history. The view back to the past and a knowledge of history is the best means of securing peace,' commented Coubertin. He ended his speech with the famous sentence: 'Demanding that nations should love each other is a childish utopia. However, the demand for mutual respect is realistic. But people must know each other before they can respect each other.'

30 *H. Schelsky,* Friede auf Zeit: Die Zukunft der Olympischen Spiele, *Texte + Thesen 30 (Osnabrück: Fromm, 1973).*

In sum, for Coubertin, Olympism was a kind of education for peace by performing Olympic sports in an Olympic setting, during the Olympic Games. The Games are symbols for a peaceful life in a modern world, which is all too full of conflict, but also of dynamism and progress.

10.2 Olympism today: The IOC and the Olympic principles

The German philosopher Hans Lenk, a former Olympic rowing champion and gold medallist (eight) of 1960, talked about the ambiguity of modern Olympism.[31] Olympism had never been a clearly defined philosophy of the Olympic Games. Rather, Olympism had to be adapted to the conditions and necessities of the presence and future. At its core, Olympism is a pedagogical idea enacted through bodily sports and performance. As with every pedagogic notion, it has to be constantly updated and modernized.

Among others, this process is not least promoted by the IOC, as head of the international Olympic institutions. Compared to the Catholic Church, the IOC can be seen as the Vatican of the worldwide Olympic sports religion. The IOC in Lausanne, Switzerland, not only manages current Olympic affairs and businesses, but also deals with its own Olympic heritage and ideological basics, similar to the department of dogmatic theology in the Vatican in Roma, Italy.

In fact, the contemporary version of Olympism is fixed in the constitution of the IOC, the so-called Olympic Charter, in other words, the bible of the Olympic Movement.[32] The latest version of the charter, revised in 2014, includes seven 'fundamental principles of Olympism'. The first paragraph argues that 'Olympism is a philosophy of life, exalting and combining in a balanced whole, the qualities of body, will and mind. Blending sport

31 *H. Lenk,* Werte, Ziele, Wirklichkeit der modernen Olympischen Spiele, *2., verb. Aufl., Beiträge zur Lehre und Forschung der Leibeserziehung 17/18 (Schorndorf bei Stuttgart: Hofmann, 1972).*

32 *www.olympic.org/.../olympic_charter_en.pdf (accessed 25/08/2015).*

with culture and education, Olympism seeks to create a way of life based on the joy of effort, the educational value of good example, social responsibility and respect for universal fundamental ethical principles.' The following six paragraphs are related to the contribution of Olympic sport for a 'peaceful society' (§ 2), for the unity of humankind and nations by international sport competitions (§ 3), and for a world 'without discrimination of any kind and in the Olympic spirit, which requires mutual understanding with a spirit of friendship, solidarity and fair play' *(§ 4)*. The last paragraphs of these still valid criteria of Olympism emphasize the autonomy and independence of Olympic sports, including the athletes themselves, as well as the Olympic sport organizations, and, finally, the promise to promote Olympic sport 'without discrimination of any kind, such as race, colour, sex, sexual orientation, language, religion, political or other opinion, national or social origin, property, birth or other status.'

It seems to me noteworthy that the IOC retains classical principles of Olympism like education through sport for a better, fair, and peaceful world without discrimination. The autonomy and independence of Olympic sport and its institutions from any political influences seem to be emphasized more than ever, perhaps by contrast to the reality of more than a hundred years of modern Olympic history, backed by the experiences of 30 Olympiads or Olympic Games. During this era, the Olympic Games were permanently in danger of being functionalized by political purposes. In addition, they are more than ever in danger of being consumed by economic constraints and interests.

The contemporary IOC Olympic Charter disclaims the notion of Olympic religion, which was Coubertin's first criterion for Olympism. Instead, the first paragraph of the charter speaks of Olympism as a 'philosophy of life'. This claim seems to be modest than the arguably blasphemous claim of Coubertin to be regarded as the founder of a new civil religion of modern Olympic sport.

To sum up a common understanding of contemporary Olympism or Olympic education – which seem to be synonymous, the common aim of Olympic education, achieved by the Olympic Games and in general by Olympic sport – is to make people strong personally and in a social context, for their own lives and that of society, community, and the state. This aim is by no means limited to top-level athletes, but relates to everyone who strives for better achievements according to the rules and principles of fair play. By doing this in the realm of sports, they improve their character in general. Top-level athletes at Olympics or Paralympics are, or at least should be, idols for all. Education, training, and competition in sport develop equally, both the whole personality and social competences like fair play, mutual respect, and – finally – the ability to live a life in peace, both with oneself and with others. This is perhaps the ultimate aim of Olympic education: to contribute toward a peaceful cooperation of individuals, teams, peoples, states, and nations, by means of competitive sports.

10.3 Olympic education with respect to universal educational principles

The term *Olympic education* encapsulates various educational opportunities created by sport in general. They are related to the ethics of Olympic sport, which find their expression in the practice of sport, and in its written and unwritten principles alike. They can be seen as the outcome of proposals and opinions put forward by various international experts in the field of Olympism and Olympic education.

Three essential elements can be regarded as relevant for universal perspectives of education in general.

10.3.1 Working on the 'self'

Firstly, there is the 'will to be better' and the longing for perfection, as the German sport-pedagogue Ommo Grupe described the anthropological essence of the idea of competition in Olympic sports.[33] This not only holds true for the outstanding performance of some athletes, but for all humans, whether or not they are able to achieve world records. The

33 *G. Ommo, "Olympische Pädagogik," in* Olympische Erziehung - eine Herausforderung an Sportpädagogik und Schulsport, *ed. Rolf Geßmann, 1st ed. (Sankt Augustin: Academia Verl., 2004).*

only thing that counts is the 'will to be better', the will to rise above oneself, even if there are obstacles in the way. The handicapped athletes who take part in the Paralympics, which have taken place directly after the official Games since Barcelona 1992, show in a most impressive manner, how one can challenge the limits set by nature, illness, accidents, or fate.

This challenge of Olympic education is rooted in the ancient, new-humanistic ideal of *Selbstvollendung* (perfection of the 'self') and *Selbstgestaltung* (creation of the 'self'), which still prevails in the form of terms like *self-determination, personality development*, or *emancipation*. 'The will to be better' also means not accepting limitations and restrictions, but fighting for changes in the fields of individuality, society, politics, economy, and culture.

This idea of progress and perfection in modern sports finds its expression in the Olympic motto 'altius, citius, fortius.' However, this saying should be seen critically from a pedagogic standpoint, as the unconditional striving for more might be at the cost of one's own health or that of partners and opponents. Therefore, any competition should be structured and restricted by sound and humane rules and norms.

The striving for being better, however, is the exceptional feature of Olympic sport, even if it is not successful in the end. True character becomes only visible when much is at stake. Those who play by the rules, give a chance to their opponents and obey the unwritten rules of fair play in the face of success and failure alike, show true Olympic spirit.

Therefore, as the philosopher Wilhelm Schmid pointed out, the field of sport is ideal for learning 'the art of life', as not only defeat and loss, which are part of every human life, but also winning, success and triumph can be experienced. In no other sphere of life are pride, arrogance, or resting on one's laurels punished as quickly as in sport.

10.3.2 Flow

Secondly, the feeling of a complete and energized focus on an activity, with a high level of enjoyment and fulfillment. This element of Olympic pedagogy is closely related to concepts that can be found in Asian, especially Japanese, philosophy and pedagogy.

Otto Friedrich Bollnow, a German philosopher and educator, whose works are widely read in Japan, speaks of the 'spirit of practice', which plays a decisive role in Japanese culture and education.[34] This spirit can not only be observed in the art of archery, but also in learning Japanese script or in the art of ikebana. Kano Jigoro was one of the prominent Japanese educators, the founder of modern Judo, who transferred the Japanese spirit of practice or exercise by concentration and contemplation to the international philosophy of Olympism.[35]

Japanese athletes are not among the most successful in the world, but in the sports where they demonstrate outstanding performance, for example in judo or gymnastics, the Japanese spirit, which has become a universal virtue of Olympic sports, is very evident.

Every athlete who wants to call himself 'Olympic' and who wants to be victorious, should learn these unique principles and practices.

The American social psychologist Mihaily Czsikszentmihalyi, who is of Hungarian ancestry, demonstrated in several publications that a complete focus on an activity, called

34 *Otto Friedrich Bollnow,* Vom Geist des Übens: Eine Rückbesinnung auf elementare didaktische Erfahrungen, *3., durchges. u. erw. Aufl. (Stäfa, CH: Edition Rolf Kugler im Rothenhäusler Verl, 1991). See also with respect to Japan Karlfried von Graf Dürckheim,* Japan und die Kultur der Stille, *4th ed. (Weilheim, Obb.: Otto Wilhelm Barth-Verl., 1949)*

35 Andreas Niehaus, Leben und Werk Kanô Jigorôs: (1860 - 1938); ein Forschungsbeitrag zur Leibeserziehung und zum Sport in Japan, *2., überarb. Aufl., Sport, Kultur und Gesellschaft 4 (Würzburg: Ergon-Verl., 2010), Univ., Diss. u.d.T.: Niehaus, Andreas: Kanô Jigorô (1860 – 1938), Köln, 2002.*

'flow', is not dependent on what one is doing or what one gets out of the activity.[36] What is more relevant is the setting of realistic goals for oneself and the constant will to achieve them. This can be learned from Olympic sports. Every athlete has to do this if he wants to be successful; if he does not, he will fail. Financial gains fade into the background in the face of this primarily intrinsic motivation.

10.3.3 Social training

Thirdly, Olympic education includes a specific and modern concept of social education or social training, namely the voluntary observance of rules, norms, and principles, extending from the written rules of sports competition to more abstract principles of equality, fairness, and modesty in the face of success and failure alike. It is extremely important for life in a democratic society that these guidelines be accepted, especially in situations when something is at stake. Fairness should not be a luxury good. The difficulties that arise when this theory is to put into practice can be observed not only in the field of sports, but also in any other aspect of life.

Coubertin referred to Olympic sport as a school for democracy, and he regarded the support of democracy and internationalism as the most important task of the Olympic Movement.

Accordingly, the term *Olympic education* defines an education for internationality and universality **in** the field of sport as well as **by** sport. As any other area of life, Olympic sport has become a symbol of universality and internationality. Even Coubertin stressed the fact that athletes from all over the world should treat each other with tolerance

36 *Mihaly Csikszentmihalyi,* Beyond boredom and anxiety: The experience of play in work and games, *2. print, The Jossey-Bass behavioral science series (San Francisco Cal.: Jossey-Bass, 1977); see also in German Mihaly Csikszentmihalyi,* Das Flow-Erlebnis: Jenseits von Angst und Langeweile: im Tun aufgehen, *8. Aufl. (Stuttgart: Klett-Cotta, 2000).*

and 'mutual respect'.[37] According to Coubertin, these meetings and peaceful contests in the field of sport contribute to peace and understanding among nations, if they are carried out in the spirit of sport and fair play. This is the central idea of Olympic education, introduced by Coubertin, and that can still be found in the Olympic Games and in Olympic sports and that has been advanced theoretically by various philosophers and pedagogues from all parts of the world. Perhaps this is also a reason for the worldwide fascination with the Games.

There is a link between this point and the pedagogic ideas of one of the most famous psychologists and social scientists of our time. The Genevan psychologist *Jean Piaget* was convinced that a modern and advanced pedagogy had to prepare children and teenagers for life in a world that is growing together, becoming international or – as we would say nowadays – global. This education for peace does not take place automatically but has to be based on psychology, namely on the developmental processes in children and young people, as *Piaget* argued.[38] According to him, these processes can only be regarded as successful if the natural egoism or egocentrism of children (or of humans in general) can be overcome and complemented with actions and thoughts marked by solidarity. As Piaget states, this cannot be achieved by an education that is characterized by obedience and lecturing – Piaget calls this 'internal solidarity' as distinguished from 'external solidarity' – but only by activity and experience, meaning 'self-government', by getting to know each other, through communication, cooperation, social performance, and combined action.

37 *Pierre de Coubertin,* Die gegenseitige Achtung, *with the assistance of Hildegard. (Müller (Sankt Augustin: Academia Verl. Richarz, 1988).*

38 *Jean Piaget,* Über Pädagogik, *Gek. Dt. Erstausg., BT/ Essay 1 (Weinheim, Basel: Beltz Taschenbuch, 1999), Aus dem Französischen von Iréne Kuhn und Ralf Stamm; mit einer Einleitung von Silvia Parrat-Dayan und Anastasia Tryphon.*

10.3.4 The role of Olympic education in the modern education system

The term *Olympic education* summarizes in a unique way the international perspective of a modern pedagogy, demanded by Piaget; a pedagogy at whose center are peace and solidarity, whose morals can be mediated only by reasoning, but also by means of feelings and actions. This pedagogy works by means and methods, which are known, popular and diffused around the world – namely sports.

Sport pedagogy in terms of Olympism does not ignore the minds and bodies of people, but addresses them with the universal language and symbolism of sport. Olympic education plays an independent and extraordinary role in the modern and international system of education. Olympic education reaches people throughout the world in the form of the Olympic Games. In some countries, Olympic education has been included in the systems of public education, as special projects have been established in schools. All of this is very important for promoting the Olympic idea and the pedagogic goals of the Olympic Movement.

The most important thing however is that the pedagogic message be transported in a credible manner by the Games themselves. It is in this respect that I see the greatest dangers for the future of Olympism. Nowadays, many people relate to the Olympic Games to a lesser extent as education than as commerce, show business, doping, television, scandals, and the like – in short, to things that have to be regarded as negative, harmful, and uneducational. Olympic sports are often perceived as elements of international commerce and of the entertainment industry and not as a means of sophisticated education and culture that is diffused globally.

Pierre de Coubertin left an important legacy as the president of the IOC, when he addressed the IOC members in his farewell speech of 1925 in Prague. Today, the words he spoke seem more important than ever. Coubertin said, 'Marketplace or temple, sportsmen have to choose. They cannot have both, but they do have to agree on one thing.

Sportsmen, make your choice!' For Coubertin this was not a dilemma. He decided in favour of the temple: 'The temple will stand forever, everything else will fade away.'[39]

No doubt, Olympism continues to be a very ambitious concept for education, culture, and the world as a whole. However, this philosophy of Olympic sport apparently contrasts with the reality of the Olympic Games and Olympic sport. The gap between ideal and reality threatens to destroy the credibility of the Olympic Movement. However, this diagnosis is not new. Olympic crises have accompanied Olympic history from the beginning, according to Willi Daume, the great German Olympic thinker, former vice president of the IOC and organizer of the Munich Olympics of 1972.

Despite scandals like doping, extreme commerce, and corruption, the popularity of modern sports and Olympics are unchallenged. 'Everyone knows that the Games are more popular than ever before, that billions of people are fascinated by them, and that the Games have achieved an incredible amount in all areas,' commented Daume in 1990, and his statement is still valid today. 'And it is for sure,' he added, 'that the Olympic Games themselves are the basis of the mysterious phenomenon known as the "Olympic idea." The Games are, however, more important than the idea.'[40]

This sentence was formulated in 1989, when the world had to face radical changes and the uncertain future of sports, of the Olympic Movement and of the Olympic Games. In the era after the Cold War, which was marked by the dynamic process of globalization, the Olympic Games became one of the few unifying elements of global culture.

Today, the Olympic Games not only are one sports event among others, they are also the most important event in the field of sports. However, the Olympic Games are more than an 'event'. They are the embodiment of a worldwide idea of modern sports. In spite of the fact that commercial advertising is not allowed in the context of the Games, companies

39 *Pierre de Coubertin,* Der olympische Gedanke: Reden und Aufsätze *(Schorndorf b. Stuttgart: Hofmann, 1967), p. 115.*

40 *Willi Daume, "Haben die Olympischen Spiele und die Olympische Idee (noch) eine Zukunft?," in* Kulturgut oder Körperkult? Sport und Sportwissenschaft im Wandel, *ed. Ommo Grupe, Studium generale (Tübingen: Attempto Verl., 1990).*

are eager to profit from the positive effects generated by the Olympic Games. Media groups pay incredible amounts of money for the rights to broadcast the Games and to report on them. Hardly any other event in the world is watched by so many people. Even though these people speak many different languages, they all comprehend the message delivered through the Games: 'Achieving one's best by means of strong commitment in a fair and rule-based competition, where no competitor is discriminated against; that is what lies at the heart of the idea.'[41]

The Games are more important than the idea behind them because actions speak louder than words; this is how one should surely understand Willi Daume's statement. The Olympic idea may be interpreted in different ways by philosophers and ideologists of all shades in all parts of the world, but its true message consists of the Games and the sports themselves. The language of the Games and of sports is international and universal. It is and it can be understood even by those who have never read a single line about the Olympic idea and who have never heard a speech on Olympism, but who simply watch and experience the Games and who are filled with enthusiasm by them.

Those who perceive Olympic athletes as examples, who practice sports themselves, and who want to act according to the Olympic ideals outside the world of sport, are inspired even more by aspects like the will to exert oneself, the will to progress, fairness toward others, giving everyone a chance, and not excluding anyone who obeys the rules.

It is very important that this message be regularly disseminated and made visible by the Games. The Games are inaugurated and closed with ceremonies. These events have developed into total art works that include music, movement, dance, and stagecraft. In the course of these events, famous artists of the host country perform, who present its art and culture. Each NOC is represented by its own flag and national anthem. The athletes take a solemn oath. During the presentation ceremonies, the victorious Olympians are awarded their medals and their respective national anthems are played in honour of them.

---

41 Ibid., p. 274.

The famous words of Avery Brundage, president of the IOC at the Olympic Games of Munich 1972, 'The Games must go on', can be understood in this way: despite all problems, despite murder and terrorism – like in Munich 1972 – despite doping and commerce, despite political constraints, despite war, crime, discrimination, and injustice, despite all the bad things of this world, even despite the weakness and blame of humans in body, mind, and character, 'the Games must go on!' And according to Willi Daume, the Games must go on because the Olympic message cannot be understood without the Games.

In fact, reality and vision are closely connected in Olympism and Olympic education. Roland Naul knows about that: as an empirical scholar of education, he is aware of the power of facts, and as a visionary physical educator, he never gave up the hope that the fundamental ideas of Olympism can pave the road to a better, truly Olympic sport for all.

References

Bollnow, O. F. (1991). *Vom Geist des Übens: Eine Rückbesinnung auf elementare didaktische Erfahrungen.* 3., durchges. u. erw. Aufl. Stäfa, CH: Edition Rolf Kugler im Rothenhäusler Verl.

Burckhardt, J. (1982). *Griechische Kulturgeschichte.* 2. Aufl., 13.-15. Tsd. Dtv 6078,4. München: Dt. Taschenbuch-Verl, (Original 1898–1902).

Csikszentmihalyi, M. (1977). *Beyond boredom and anxiety: The experience of play in work and games.* 2. print. The Jossey-Bass behavioral science series. San Francisco, CA: Jossey-Bass.

Csikszentmihalyi, M. (2000) *Das Flow-Erlebnis: Jenseits von Angst und Langeweile: im Tun aufgehen.* 8. Aufl. Stuttgart: Klett-Cotta.

Daume, W. (1990). "Haben die Olympischen Spiele und die Olympische Idee (noch) eine Zukunft?" In O. Grupe (Ed.), *Kulturgut oder Körperkult? Sport und Sportwissenschaft im Wandel* (pp. 273–288). Studium generale. Tübingen: Attempto Verl.

De Coubertin, P. (1967). *Der olympische Gedanke: Reden und Aufsätze.* Schorndorf: Hofmann.

De Coubertin, P. (1974a). *Einundzwanzig Jahre Sportkampagne (1887–1908): (Une campagne de vingt-et-un-ans).* With the assistance of B. Wirkus. Ratingen, Kastellaun, Düsseldorf: A. Henn Verlag.

De Coubertin, P. (1974b). *Einundzwanzig Jahre Sportkampagne 1887–1908: (Une campagne de vingt-et-un ans).* Düsseldorf: Henn.

De Coubertin, P. (1988). *Die gegenseitige Achtung.* With the assistance of Hildegard (Müller). Sankt Augustin: Academia Verl. Richarz.

De Coubertin, P. (1996). *Olympische Erinnerungen.* Reprint der 2. Aufl. 1959. Wiesbaden: Limpert.

De Coubertin, P., & Organisationskomitee für die XI. Olympiade Berlin 1936 e.V. (1935). *Pax Olympica: Weltsendung des Reichssenders Berlin am Sonntag, dem 4. August 1935 mittags.* With the assistance of Pierre d. Coubertin and Carl Diem. Berlin-Charlottenburg: Organisationskomitee für die XI. Olympiade Berlin 1936 e.V. [Vorolympische Kampagne in drei Sprachen].

Von Graf Dürckheim, K. (1949). *Japan und die Kultur der Stille* (4th ed.). Weilheim, Obb.: Otto Wilhelm Barth-Verl.

Hübner, Emanuel. (2015) *Das Olympische Dorf von 1936: Planung, Bau und Nutzungsgeschichte.* Aufl. 2015. Paderborn: Schöningh Paderborn.

Lenk, H. (1972). *Werte, Ziele, Wirklichkeit der modernen Olympischen Spiele.* 2., verb. Aufl. Beiträge zur Lehre und Forschung der Leibeserziehung 17/18. Schorndorf bei Stuttgart: Hofmann.

Müller, N. (Ed.). (2000). *Pierre de Coubertin, 1863–1937: Olympism.* Lausanne: IOC; Selected Writings.

Niehaus, A. (2002). *Leben und Werk Kanô Jigorôs: (1860–1938); ein Forschungsbeitrag zur Leibeserziehung und zum Sport in Japan. 2.,* überarb. Aufl. Sport, Kultur und Gesellschaft 4. Würzburg: Ergon-Verl., 2010. Univ., Diss. u.d.T.: Niehaus, Andreas: Kanô Jigorô (1860–1938): Köln.

Nietzsche, F. W., & Acampora, C. D. (1996). *Homer's contest.* Nietzscheana/North American Nietzsche Society 5. Urbana, IL: North American Nietzsche Society.

Ommo, G. (2004). "Olympische Pädagogik." In R. Geßmann (Ed.), *Olympische Erziehung – eine Herausforderung an Sportpädagogik und Schulsport* (1st ed., pp. 35–52). Sankt Augustin: Academia Verlag.

Piaget, J. (1999). *Über Pädagogik*. Gek. Dt. Erstausg. BT/Essay 1. Weinheim, Basel: Beltz Taschenbuch, 1999. Aus dem Französischen von Iréne Kuhn und Ralf Stamm; mit einer Einleitung von Silvia Parrat-Dayan und Anastasia Tryphon.

Rittberger, V., & Boekle, H. (1997)."Das Internationale Olympische Komitee – eine Weltregierung des Sports?" In O. Grupe (Ed.), *Olympischer Sport: Rückblick und Perspektiven* (pp. 127–156). Schorndorf: Hofmann.

Schelsky, H. (1973) *Friede auf Zeit: Die Zukunft der Olympischen Spiele,* Texte + Thesen 30. Osnabrück: Fromm.

PART FOUR:

PHYSICAL ACTIVITY AND HEALTH: HEPA POLICY AND DEVELOPMENT IN THE EUROPEAN UNION

11 The challenge of HEPA for physical education in Europe

Claude Scheuer

11.1 Introduction

One of the priorities on the political agenda related to sport in the EU over the past years has been health-enhancing physical activity (HEPA). This political priority impacts also physical education (PE) as the only school subject where all students are physically active, in a sense that international recommendations on daily physical activity can at least be partly reached. In turn, European associations focusing on PE face challenges to address the priority of HEPA as part of their respective missions. However, HEPA can be considered as a chance for these associations to raise their profile and the profile of PE as a school subject when accentuating their potential to promote HEPA in school settings.

In the following, this chapter will focus on HEPA as a relevant political priority for PE in Europe, before focusing on two European associations active in the field of PE: the European Physical Education Association (EUPEA) and the European Council for Physical Education and Youth Sport (CEREPS, *Conseil Européen des Recherches en Éducation Physique et Sportive*). Roland Naul is an individual member of EUPEA and a founding member of CEREPS. Furthermore, the Willibald-Gebhardt Institute is an institution member of both EUPEA and CEREPS.

11.2 HEPA policy relevant for PE in Europe

In the past years, HEPA has turned into a priority of EU policy in sport. Several policy documents addressing physical activity, which are listed below, have been issued by different EU institutions:

- Study on young people's lifestyles and sedentariness and the role of sport in the context of education (2004)
- White Paper on Sport (2007)
- EU Physical Activity Guidelines (2008)
- EU competence for sport (Art. 165 TFEU; 2009)
- Preparatory actions in the field of sports (2009–2013)
- Communication on sport (2011)
- EU Work Plan for Sport (2011–2014)
- European Parliament Resolution 'European dimension in sport' (2012)
- Council Recommendation on HEPA (2013)
- Erasmus+: Sport programme (2014–2020)
- EU Work Plan for Sport (2014–2017)
- EUC Expert Group 2014–2017 on HEPA: Recommendations to encourage PE in schools, including motor skills in early childhood, and to create valuable interactions with the sport sector, local authorities and the private sector
- EU Work Plan for Sport (2017–2020)

Especially the recommendations to encourage PE in schools of the Expert Group 2014-2017 on HEPA of the European Commission is of special interest for a European association focusing on the promotion of PE, as it is the case for EUPEA or CEREPS. Below are listed the 10 priorities that the 28 recommendations focus on (EUC Expert Group on Health-enhancing physical activity, 2015).

- physical literacy and fundamental motor skills to be promoted before any school setting
- PE curricular content
- Inclusive approach
- Injury prevention
- PE taught time

- Exemptions from PE
- Assessment in PE
- PE teachers
- Monitoring of PE
- Extracurricular activities and activities outside PE curriculum

In the following, some particular relevant recommendations on PE curricular content, inclusion, PE time allocation, PE teacher education and extracurricular physical activity are listed.

Priority 'Physical education curricular content'

> *Recommendation 6 – 'Physical education and extra-curricular activities should foster an ethical education by teaching values such as fair play, cooperation, equity, equality, integrity, peace, human rights, and respect of others' capabilities. Through sport participation, they should also develop relevant skills such as teamwork, social inclusion and leadership, avoiding sport stereotypes.'*
>
> *EUC Expert Group on Health-Enhancing Physical Activity, 2015, p. 9*

> *Recommendation 7 – 'The physical education curriculum should include health education concepts like personal and social well-being, health promotion, and healthy lifestyles from a broader perspective beyond the practice of physical activity and sport. Physical education teachers should also cooperate closely with other disciplines in school to fully develop these concepts among the education community.'*
>
> *EUC Expert Group on Health-Enhancing Physical Activity, 2015, p. 10*

Priority 'Inclusive approach'

> *Recommendation 8 – 'Everyone should be able to participate in physical education and extra-curricular activities through inclusive, differentiated and adapted methodologies and activities, including less active and less skilled children. Children with a disability or special educational needs should be offered adapted activities and not be excluded.'*
>
> *EUC Expert Group on Health-Enhancing Physical Activity, 2015, p. 10*

Priority 'Physical education taught time'

> *Recommendation 10 – 'The minimum physical education taught time recommended during compulsory education period should be increased to at least 5 lessons per week (~ 5 hours). The physical education curricular structure and goals should be adjusted accordingly, defining tangible and flexible outcomes for each developmental stage, and suggesting the inclusion of realistic activities.'*
>
> *EUC Expert Group on Health-Enhancing Physical Activity, 2015, p. 13*

Priority 'Physical education teachers'

> *Recommendation 13 – 'Qualified and specialised PE teachers should be preferred at all educational levels. When not possible, as a minimum, qualified PE teachers or certified coaches should counsel and support general teachers.'*
>
> *EUC Expert Group on Health-Enhancing Physical Activity, 2015, p. 14*

Priority 'Extra-curricular activities and activities outside physical education curriculum'

> *Recommendation 16 – 'Schools, alone or accompanied by other relevant organisations should promote and increase the availability of physical activities outside physical education curriculum (e.g. physical activity and sport, active breaks) including the implementation of the active school concept.'*
>
> *EUC Expert Group on Health-Enhancing Physical Activity, 2015, p. 16*

As outlined in these policy documents, physical activity in general and HEPA in specific are of primordial importance to tackle the current physical inactivity crisis in Europe and beyond. That said, it is a challenge for all stakeholders active in all relevant settings to promote HEPA. When it comes to school settings, associations and organizations promoting PE and physical activity in general in this setting of highest importance will find opportunities to align their mission to the political priorities described above. In the following, two European associations active in this field will be presented.

11.3 Activities of and challenges for the EUPEA related to HEPA

The EUPEA was founded in 1991 in Brussels (Figure 11.1). EUPEA is the umbrella organization of the national Physical Education Associations in Europe. As a non-profit and non-governmental professional organization of national PE Associations, EUPEA promotes PE and school sport in Europe by working and cooperating with appropriate governmental and non-governmental organizations and individual experts and professionals. The mission of EUPEA was defined in the Declaration of Madrid in 1991 (amended in Brussels in 2011) and turns around the main slogan 'No Education without Physical Education' (EUPEA, 1991). Thus, the focus of EUPEA lies on the promotion of and the advocacy for PE in Europe. This is done in cooperating with other international agents and associations, like ICSSPE (*International Council for Sport Science and Physical Education*), FIEP (*Fédération International en Éducation Physique*), AIESEP (*Association Internationale des Écoles Supérieures d'Éducation Physique*) or CEREPS (*Conseil Européen des Recherches en Éducation Physique et Sportive*). Furthermore, EUPEA has contact with the European Union Commission and the Council of Europe as observer in Expert Groups of the EUC and as a member in the Consultative Committee of EPAS (*Enlarged Partial Agreement on Sports* of the Council of Europe).

Figure 11.1: EUPEA logo

Besides the focus on PE in general, several EUPEA documents highlight as well the importance of HEPA. In the following, relevant statements regarding HEPA in these documents will be presented.

Declaration of Madrid 'No Education without Physical Education', 27th of October 1991, amended 10 November 2011 in Brussels by the 'add that' points (EUPEA, 1991)

The following paragraphs in the *EUPEA Declaration of Madrid* refer to HEPA:

7. *There is a need to emphasize the holistic approach to Physical Education as a subject for health empowerment and teaching/learning values and ethics.*

8. *Physical Education makes a major contribution to the motor, cognitive and social development of children and is one of the most important bases of lifelong physical activity.*

9. *Cooperation between schools, sports clubs and the local community must be enhanced and the coordinating role should be part of the PE teacher's job.*

EUPEA Physical Education Guidelines (EUPEA, 2000)

In the *EUPEA Physical Education Guidelines*, the following statements are linked to HEPA as well:

5. *Investment in the young through PE promotes active lifestyles and health throughout life.*

6. *In order to give all pupils a comprehensive experience in PE, physical activity and sport, schools should work with partners outside the school and make use of all the opportunities in the local environment.*

7. *More opportunities should be offered to students in further and higher education to gain credits for engagement in physical activity.*

EUPEA Principles (EUPEA, 2001)

Also, the *EUPEA Principles* include references that can be linked with HEPA:

1. *Ensure that all pupils, regardless of ability or disability, culture, religion, gender or social class have access to a high quality physical education curriculum.*

2. *Promote a broad and balanced physical education curriculum that encompasses the skills, knowledge and understanding to ensure that children and young people are physically educated.*

9. *Recognize the contribution of physical education within further and higher education and its lifelong commitment to health, physical activity and the National sporting culture.*

EUPEA research on a physically well-educated person

In a study on how a physically well-educated person should look, more than 700 PE teachers from 14 European countries were asked about the main aims of PE (Fisher, Repond, & Diniz, 2011). In this comparison from priorities adopted by PE teachers, the following aims were ranked in the top five, with two HEPA-related aims in the two leading positions:

5. *To encourage the adoption of active and healthy lifestyles*

6. *To develop a feeling of personal well-being/wellness*

7. *To inculcate a sense of important values in sport (fair play, solidarity . . .)*

8. *To develop a broad repertoire of movement competence*

9. *To provide opportunities for all pupils to learn and achieve, regardless of ability, gender or social and cultural background*

The aims and learning outcomes prioritized by teachers demonstrate that the important factors for them are the adaption of a balanced and active life with good habits. In general, the statements found in official EUPEA documents underline that the umbrella association for PE teachers in Europe has identified HEPA as a priority in the frame of its mission to promote PE in Europe for many years already. This conclusion can be identified as well, when analyzing the priorities of the European projects EUPEA is and was involved in during recent years (see also the chapter on *Erasmus+-funded HEPA-promoting projects in Europe: Project activities of the European Physical Education Association* in this book).

11.4 CEREPS, a European council for research in physical education and youth sport[1]

CEREPS was founded as a non-governmental organization in accordance with Luxembourgish law (a.s.b.l.) with support of the Willibald Gebhardt Institute (WGI) in Luxembourg in 2015 (Figure 11.2). The council organized its first meeting in 2016 at the University of Luxembourg with researchers from more than 15 EU-countries representing different universities and EU-umbrella organizations, followed by the CEREPS Summit of 2017 in Brussels organized by the University of Münster (WWU) and the WGI at the Office of the State of North-Rhine Westphalia at the European Commission.

Figure 11.2: CEREPS logo

11.4.1 Why CEREPS?

Research about implementation efforts and measurement outcomes of PE, school sports and any other physical activity, whether it is in curricular time at school or extracurricular time outside school, for example, in sport clubs, needs to be more focused on different levels today. Children and adolescents, their parents, teachers and coaches need to be informed about what is necessary for the individual support of our children and young people to stay or to become more physically active not only for various health reasons, but also for a well-rounded education with psychosocial qualities and cognitive benefits. Evidence-based research outcome about quality in PE teaching and boys' and girls' daily

1 *This text is based on a collaborative contribution from Claude Scheuer and Roland Naul to the 12th FIEP European Congress 2017 in Luxembourg (Scheuer & Naul 2018).*

involvement in various forms of organized and unorganized physical activities help children and adolescents to achieve better life skills, to get socially included and be respected by their peers, which often go alongside with more personal self-esteem and self-efficacy and also better learning outcomes at school. However, many national and international associations in the field of PE, school and youth sport based in Europe do not have a special council or committee dedicated to research, monitoring or evaluation of teaching and learning in PE and school sports. In particular, national and international youth sport organizations in Europe working on grass-root or elite level face difficulties in linking their offers and interventions with monitoring and evaluation of their many activities and projects if there are no scientific consultants or any academic partners for support and assessment of the outcome. But academic institutions and scientific associations also need to verify their scientific concepts and theoretical assumptions through practical applications about their promised impacts in related settings. In recent years, some stakeholders who set up projects in PE, school and youth sports are looking even more to these 'collaborative partnerships' and vice versa to link practice and theory and to bridge the gap between implementation, diffusion and evaluation through linked partnerships on the national, cross-national and EU-level.

CEREPS as a research-related European council picks up these new 'partnership developments' across countries in Europe, bringing together different partners, institutions and umbrella-organizations out of the fields of practice and theory in PE, school and youth sport. We agree on the slogan: There is no good education without PE; however, there is no good PE without evidence-based documentation about the process and outcome of their aims and objectives. Today, the promotion of PE and school sports in terms of quality and quantity needs evidence-based research results to raise funding and achieve more awareness and better support through national and European-based policymakers. CEREPS, as the European Council for Research in Physical Education and Youth Sport, will link with all national and EU-umbrella organizations in the field of PE, physical activity and grassroots sports for children and adolescents, in order to assist with common research efforts to get more visibility and support inside our common sectors of education, health and sports across Europe for the benefits of our future active young Europeans. New local, regional and national concepts and strategies of PE, school and youth sport need better exchange, stronger common ties and comparable research items between EU-countries.

11.4.2 What is CEREPS?

CEREPS is a non-governmental, non-profit organization under Luxembourg law (a.s.b.l.) and offers individual and institutional memberships as a body of coordination of research items and cross-national projects in all PE and physical activity settings for children and adolescents. CEREPS wants to become a major address for research development and applications. It already serves for its members and partner organizations across Europe in the capacity of a stock exchange on research items of PE, school and youth sport including health benefits of regular physical activities at inter-institutional and inter-country levels.

The range of CEREPS' body of knowledge with related research items covers four main settings of physical activity at school, where also network partnerships exist between local community authorities in the sectors of education, health, social work and sport. CEREPS' cross-sectoral partnerships in research include also sport organizations from local, regional, national up to EU level. Figure 11.3 shows the four main settings CEREPS is focusing on:

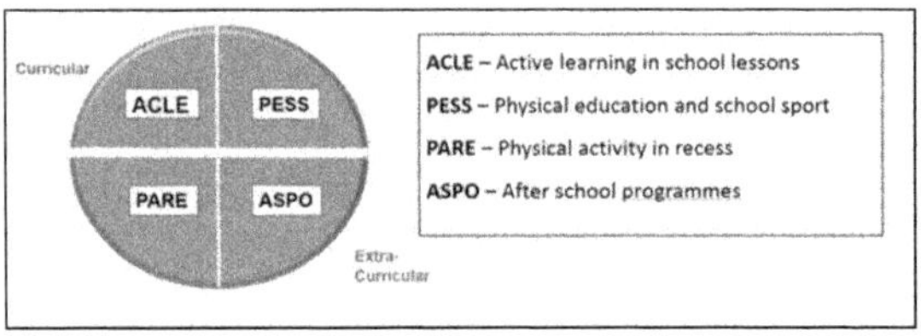

Figure 11.3: The four physical activity settings in schools (Scheuer & Naul, 2018, p. 66)

There are two curricular settings: active learning with physical activities in different school subjects, like languages, mathematics, natural sciences and others (ACLE). Another important setting is the subject of PE including school sport, in some countries as a traditional or as a new part of the physical education lesson (PESS).

There are also two extracurricular settings: physical activity between lessons and as a part of regular breaks in school life, both in so-called recess time at school (PARE), where school sports in some countries are also included. Another setting, which gains more and more attention across Europe via new school laws and new school reforms is the after-school programme with a variety of sports and physical activities, often organized by the school with support of local partners, mainly from the sport and health sector (ASPO).

CEREPS works as a Europe-based council to enhance research on and networking for these two curricular and extracurricular settings for active living of children and adolescents. The mission of CEREPS is to promote the development of age- and gender-related good competencies in fundamental motor skills with psychosocial qualities and cognitive benefits as a holistic approach of a well-rounded individual with basic life skills via research-based documents.

11.5 Outlook

The topic of HEPA is a priority when it comes to the implementation of EU sport policy and is very likely to remain a priority in the future, as the related challenges can only be addressed on long term. This makes it even more important for European PE associations to reflect on their potential to contribute to the improvement of the situation by promoting PE and physical activity in school contexts. The missions and activities of EUPEA and CEREPS presented in this chapter show their immense potential in this field. As EUPEA is representing the PE teacher profession and thus the PE practitioners active in schools, CEREPS is bringing together researchers focusing on school settings where physical activity is happening. As their work is complementary, their cooperation will be important to achieve their common goals. Together with other European and international stakeholders, as, for example, FIEP, AIESEP or ICSSPE, they certainly will play a major role in the future when it comes to addressing the challenge of HEPA in PE.

References

EUC Expert Group on Health-enhancing physical activity. (2015). *Recommendations to encourage physical education in schools, including motor skills in early childhood, and to create valuable interactions with the sport sector, local authorities and the private sector.* Retrieved from http://ec.europa.eu/transparency/regexpert/index.cfm?do= groupDetail.groupDetailDoc&id=19860&no=1

European Physical Education Association. (1991). Declaration of Madrid "No Education without Physical Education" 27th of October 1991, amended 10 November 2011 in Brussels by the "add that" points. Retrieved from www.eupea.com/public/uploads/files/documents/Declaration%20Madrid%201991%20-%20add%20that%20Brussels%202011.pdf

European Physical Education Association. (2000). Physical education guidelines. Retrieved from www.eupea.com/public/uploads/files/documents/EUPEA%20Physical%20Education%2010%20guidelines.pdf

European Physical Education Association. (2001). EUPEA principles. Retrieved from eupea.com/public/uploads/files/documents/EUPEAprinciples.pdf

Fisher, R., Repond, R.-M., & Diniz, J. (2011). A physically educated person. In K. Hardman & K. Green (Eds.), *Contemporary issues in physical education: International perspectives* (pp. 69–89). Aachen: Meyer & Meyer.

Scheuer, C., & Naul, R. (2018). The vision and mission of CEREPS, a European council of research in physical education and physical activity. In C. Scheuer, A. Bund, & M. Holzweg (Eds.), *Changes in childhood and adolescence: Current challenges for physical education. Keynotes, invited symposia and selected contributions of the 12th FIEP European Congress* (pp. 64–66). Berlin: Logos.

12 Past, current and future work of the WGI and its fit with HEPA policy and trends of PE development[2]

Dennis Dreiskämper & Till Utesch

12.1 Introduction

The Willibald Gebhardt Institute (WGI) can be seen as the oeuvre of Prof. Dr. Dr. h.c. Roland Naul in a nutshell. All his research interests, political and administrative endeavours and relations, research projects and collaborations as well as strategic efforts and influences are bundled up in the WGI. By that, the WGI is described as a caretaker of 'values in sport and education' (www.wgi.de). But what is meant by these values? When we define the different research pathways of the WGI, we can distinguish at least between four different pillars: health, youth sport, elite performance sport and sport history/ Olympic education. In the area of health, the term *health-enhancing physical activity* (HEPA) stands for the worldwide effort to foster physical activity in all age groups and all parts of the population in order to establish a healthier lifestyle. In HEPA, the 'values' were defined by the EU Physical Activity Guidelines by the EU working group 'Sport & Health' in September 2008 for the first time (EU, 2008). In this document, the EU recommends policy action in support of HEPA. Overall, 41 guidelines for action that refer to political stakeholders, public authorities and public grants are formulated. These 41

2 *This article is a textualization of the keynote speech delivered in Prague, 30 January 2017, in the ceremony for conferring a doctorate degree, honoris causa, on Prof. Dr. Dr. h.c. Roland Naul by the first author.*

guidelines can be interpreted as the 'values' that the WGI tries to follow in its various research projects and publications regarding HEPA. Subsequently, the aim of this chapter is to show how the HEPA guidelines are implemented in the work of the WGI and how the guidelines and strategy of the WGI influence each other reciprocally. By this, we will refer to the projects of the WGI that have been implemented and conducted in the last 15 years. The broad and intensive work of the WGI in terms of HEPA policy can be seen in the many research approaches in that time. Starting with the evaluation of the open all-day schools (SpOGATA, 2003–2011; Forschungsgruppe SpOGATA, 2015) and the EU-wide preparatory action in Healthy Children in Sound Communities (2008–2011; Dreiskämper & Naul, 2011), the most important project, 'Gesunde Kinder in Gesunden Kommunen' was started in 2007. Today, this project can be divided in four parts: the EU-funded multicomponent intervention project was conducted between 2007 and 2014 in 39 primary schools in Germany and the Netherlands (Naul, Schmelt, Dreiskaemper, Hofmann, & l'Hoir, 2012). Several publications that contributed significantly to the research field of HEPA and motor development have been produced as the result of this project (e.g., Dreiskaemper & Naul, 2014; Dreiskaemper, Naul et al., 2018, Naul et al., 2012; Magro, Utesch, Dreiskämper, & Wagner, 2018; Utesch, Dreiskämper, Naul, & Geukes, 2018; Utesch, Dreiskämper, Strauss, & Naul, 2018;). The idea and structure of (Healthy Children in a Sound Community) GKGK was also successfully transferred to the Chinese context (Shen et al., 2015; 2016). After ending of the EU-funding, GKGK was continued and enlarged in Dutch municipalities next to the Dutch-German border. Until today, more than 4,000 Dutch children have participated in the community-based intervention programme that aims to foster daily physical activity in primary school-aged children. Finally, GKGK was also transferred to the eastern European area and has been successfully implemented in eastern Germany (Brandenburg) and western Polish communities since 2017, again funded by Interreg. With GKGK, the WGI became a research institution acknowledged worldwide in the area of HEPA. Therefore, several projects followed within the last years. In 2015 and 2016, the WGI evaluated the German youth sport federation initiative Zukunftsinvestition: Entwicklung jungen Engagements im Sport! (ZI:EL) that aimed to reach special population groups with physical activity offers. Besides, the WGI contributed to the third German Children and Youth Sport report in 2015. In 2015, the project Healthy Growing funded by the German ministry of education was started. The WGI became a collaborator next to the German Sports University and other institutions in order to analyze conditions and predictors for healthy growing in early childhood (i.e.,

Kindergarten and primary schools). Another project, also focusing on HEPA in children is the Erasmus+ programme Active School Communities (ASC). Here, the WGI had the task to contribute to a Europe-wide handbook and a toolkit for stakeholders and practice partners. Also in 2018, the European Physical Education Observatory project (EuPEO), headed by the University of Lisboa was established. Here, the WGI is responsible for the German implementation and had an important influence on the structure and content of the project. Summarizing these projects, which cover only a part of the variety of actions of the WGI, it is obvious that the main task of the WGI is to tackle these recommendations made by the EU regarding the enhancement of physical activity. How the recommendations are considered in the projects and the practical work will be analyzed on the basis of the recommendations step by step in the next paragraph.

12.2 WGI projects and EU guidelines

The EU guidelines for action can be differentiated into different categories, either for special subgroups of the population or special stakeholders or specific recommendations for different parts of society. However, 15 of them are directly referred to in the tasks an institution like the WGI is responsible for.

Note

Guideline 1: In accordance with the guidance documents of the World Health Organization, the European Union and its Member States recommend a minimum of 60 minutes of daily moderate-intensity physical activity for children and young people and a minimum of 30 minutes of daily moderate-intensity physical activity for adults including seniors.

This, perhaps most important, guideline is referred to as the general aim of the HEPA approach. This guideline is not the only one recommendation for daily physical activity as also WHO and national references are published all over the world. For WGI projects, however, this recommendation is crucial in terms of being the centre of all intensive efforts. Especially, in GKGK the recommendation of the EU was tackled by implementing daily physical activity in primary schools (cf. Naul et al., 2012, Dreiskämper & Naul, 2014). The idea of daily physical activity habits was even transferred to non-EU countries like China. With exemplary time schedules for pupils and cooperation between schools, municipalities and sport clubs, this guideline was fulfilled for more than 1,000 pupils for the complete time period of primary school.

In other projects, the focus was on monitoring these daily physical activity times. For example, in Healthy Growing (Dreiskaemper et al., 2018) the daily habits of parents and children and kindergarten and primary schools were compared. In the project ASC a toolkit to help educators was implemented that focused on the possibility of reaching the recommended activity time per day (Naul, Utesch, & Niehues, 2018). In the project "Sport im offenen Ganztag" (SpOGATA; Naul et al., 2015, Heemsoth et al., 2015) it was analyzed whether differences in activity time could be found between children taking part in all day school offers and those who didn't. Results indicate, that all day school offers are helpful for inactive children as there are no big differences between those who are active in school and those who are not (because they may be already in a sport club, for example). Also, in the Erasmus+-funded project EuPEO, it was investigated whether differences between countries can be found in terms of school politics to stick to this particular guideline.

Note

Guideline 2 – All relevant actors should refer to the guidance documents of the World Health Organization regarding obesity and physical activity and seek ways to implement them.

GKGK emphasized specially on a multicomponent approach to reach the guideline 1 recommendation. Therefore, GKGK aimed to identify relevant actors first and brought them together. Hereby, synergies between different actors (e.g., school, municipality, sport club, as well as parents, physicians, health specialists) were produced. However, in other projects (e.g., Healthy Growing) it was found that on all dimensions, meta-, meso-, and micro- factors and actors can be found that are responsible for health behaviour. Therefore, recent results show that also specific environmental factors should be considered. Multiple actors in a community must work together to achieve some health enhancement in families and especially in children (cf. figure 12.1).

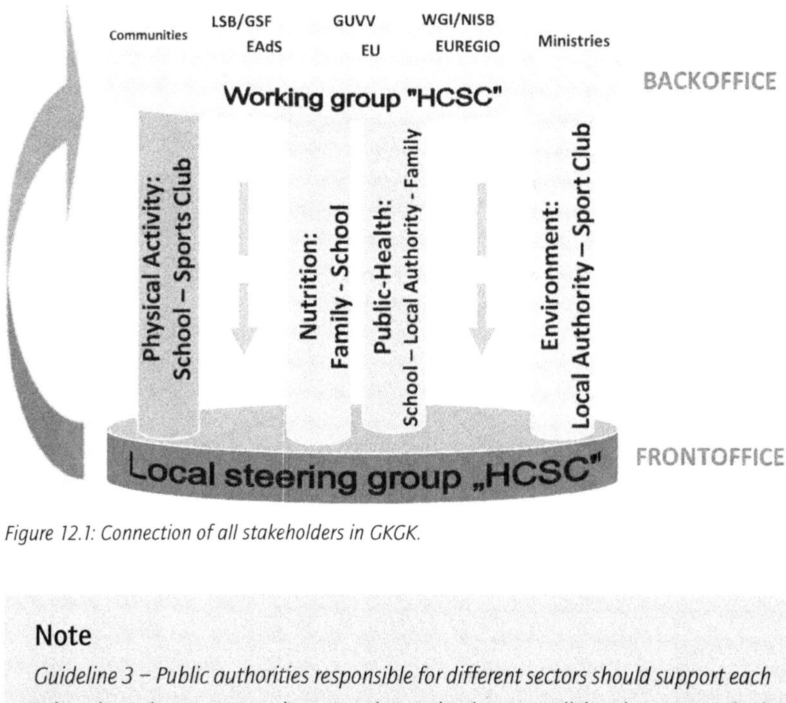

Figure 12.1: Connection of all stakeholders in GKGK.

Note

Guideline 3 – Public authorities responsible for different sectors should support each other through cross-sectoral cooperation to implement policies that can make it easier and more attractive for individuals to increase their level of physical activity.

In GKGK, the focus was on bringing different sectors of policy together in order to foster physical activity of children. The implementation of community level led to a connective work of different sectors like health, sport, youth and school policy. Besides, health insurances, physicians, sport organizations and other players were connected with each other and supported the project (cf. figure 12.2). In the project Healthy Growing funded by the

German Ministry of Education and Science (BMBF), qualitative interviews underlined that these connections are the most needed factor for successful work in the different fields of health interventions that are not established sufficiently yet (Ehlen et al., 2018). Similar results were found in the evaluation of ZI:EL (Naul et al., 2016).

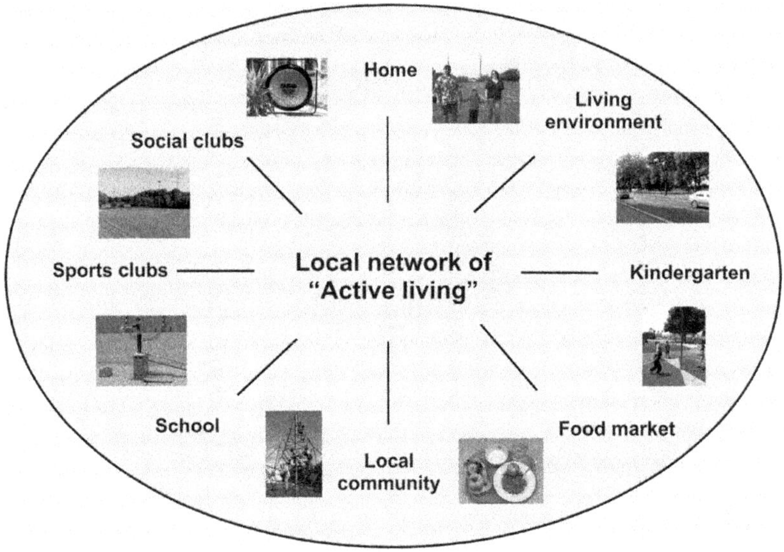

Figure 12.2: Network in GKGK.

Note

Guideline 4 – Authorities responsible for the implementation of sport and physical activity guidelines should consider the use of agreements between central, regional and local levels of Government to promote sport and physical activity. Where appropriate, such agreements can involve specific reward mechanisms. Links between sport and physical activity strategies should be encouraged.

The work of the WGI concentrated next on research on the occurrence of agreements between different stakeholders. A tremendous task was, and still is, to bring central policies down to practice (e.g., by research projects like HCSC, Healthy Children in Sound Communities) and to spread the results of regional projects nationwide (e.g., like in the Netherlands where GKGK has become one of the best-practice projects of the NISB for promoting physical activity). One of the factors, which makes HCSC and GKGK successful intervention projects, was the institution of a moderator between the different partners and levels (fig. 12.3).

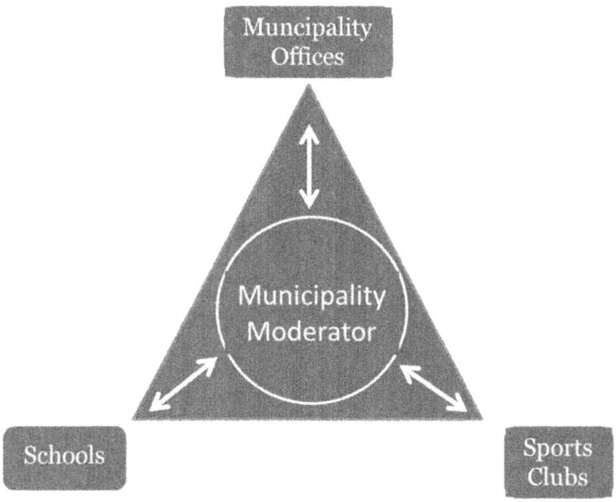

Figure 12.3: The idea of a moderator in GKGK to connect different levels and partners within an intervention.

How difficult the transfer of national or even international guidelines on a local level can be has been shown by the WGI in the evaluation of ZI:EL: the efforts to foster physical activity in the special interest groups (a) people with disabilities, (b) migrated people and (c) people with lower educational background are still the biggest challenges of a practical level that have to be fostered by multicomponent intervention studies. On an international level, the currently running WGI project EuPEO shows how difficult the standardization of different national policies is, for example, in this case in the physical education setting: a comparison between status quos, standards and procedures in several European countries emphasizes the need for international agreements on common strategies for physical activity in schools.

Note

Guideline 5 – Governments should launch initiatives to coordinate and promote public and private funding devoted to physical activity and to facilitate access for the whole population.

Still, a central aim of the WGI is to get governments more involved in projects enhancing physical activity. However, central projects of WGI like GKGK (EU), Healthy Growing (Ministry of Education and Science), EuPEO (European Commission) and SpOGATA (State North-Rhine-Westphalia) have been supported by regional, national and international governmental institutions.

Note

Guideline 9 – Sport policy should be evidence-based and public funding for sport science should encourage research that seeks to uncover new knowledge about activities that allow the population at large to be physically active.

The perhaps most important guideline that the WGI sticks to is the necessity of evidence-based science in physical activity research. It can be seen as one of the most important motivations of the WGI to deliver high-quality research in field of HEPA. Several publications during the last ten years (e.g., Forschungsgruppe SPOGATA, 2015, Naul et al., 2012, Dreiskaemper & Naul, 2014, Utesch, Dreiskämper, Naul et al., 2018) show that the aim of the WGI is not only to foster physical activity in the different projects, but to analyze how the enhancement of motor performance and physical fitness can work in general and how it is related to health benefits (cf. figure 12.4).

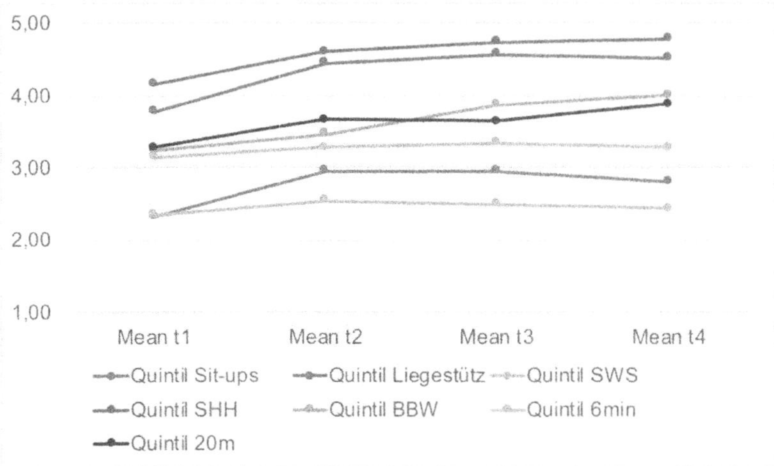

Figure 12.4: The significant improvement of physical fitness domains in GKGK within a four-year intervention period.

Note

Guideline 14 – Physical activity data should be included in health monitoring systems at national level.

In 2014, the WGI implemented the GKGK databank, by which data of physical fitness development of school children can be observed and monitored longitudinally (Naul, Utesch, & Dreiskämper, 2018, Utesch & Naul, 2015). The database serves as an important resource of information for policy and practitioners. As it was implemented in several studies and programmes within the last five years, more than 20,000 data points are stored in the databank by now. It has been translated into Dutch and Polish and is actually used for a European funded HCSC implementation in the German-Polish border area. Besides, several teachers, stakeholders and caregivers have been educated for usage and application of the databank by researchers of the WGI. The Dutch sport federation progresses in acknowledging the databank as a best practice example for monitoring and intervention support. Figure 12.5 shows how the databank is to be implemented and used in school-settings.

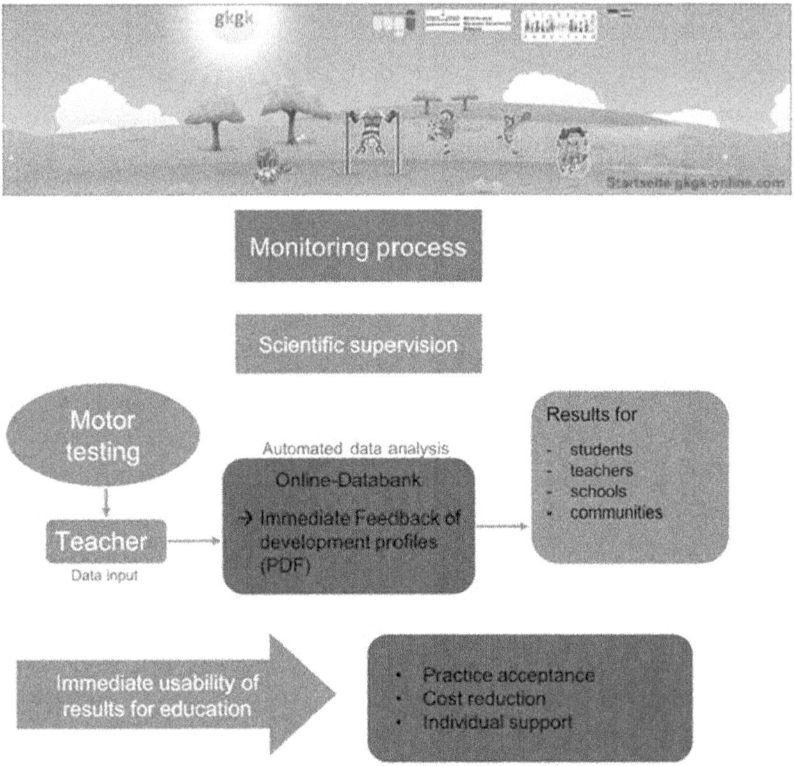

Figure 12.5: The GKGK databank.

Note

Guideline 21 – EU Member States should collect, summarise and evaluate national guidelines for physical activity addressed to physical education teachers and other actors in the development of children and youth.

Guideline 22 – As a second step, EU Member States could design health-enhancing physical education modules for the training of teachers in, respectively, kindergartens, primary schools and secondary schools.

Guideline 23 – Information about the need for physical activity, the best way to introduce it in everyday life and changes in lifestyle should be available to physical education teachers, health professionals, trainers, managers of sport and leisure centres and media professionals in the course of their studies and/or professional training.

Guideline 24 – Topics related to physical activity, health promotion and sports medicine should be integrated into the curricula of health professions in the EU.

Guidelines 21–24 are referred to within the EuPEO and ASC project, which are still currently running funded by Erasmus+. While in EuPEO international and European standards of teacher education, PE quality and PE curricula are analyzed by a multi-country-based consortium, ASC published a toolkit that provides guidelines regarding a successful implementation of a network between schools and sport clubs. These networks shall then focus on a close cooperation to enhance children's physical activity time per day. The three newly developed toolkits have been evaluated by the WGI from the perspective of physical education teachers, coaches from sport clubs and national coordinators in order to provide constructive recommendations for improvements to the toolkit (Naul, Utesch, & Niehues).

Note

Guideline 38 – The implementation of policy actions for physical activity promotion should be monitored regularly, based on pre-defined indicators to allow for evaluation and review.

The projects of the WGI focus on either implantation of intervention programmes or deriving recommendations for policy actions in different domains. Therefore, the projects are planned as multicomponent approaches fostering healthy growing and living

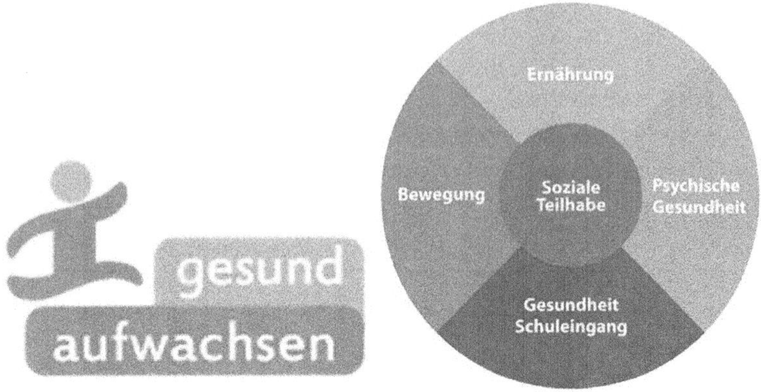

Figure 12.6: Health indicators in the project Healthy Growing.

(cf. figure 12.6). Physical activity is the central key element, but normally not the only issue addressed in the projects. For example, the interdisciplinary project Healthy Growing aims to combine the domains nutrition, psychological health, health status at start of school and physical activity in a holistic approach. The social participation in health offerings is a central key determinant that is especially related to socioeconomic factors like education background and household income. In Kindergarten and primary schools, children's health status is analyzed to find recommendations for stakeholders and politics for fostering a healthy lifestyle and environment for children and their parents. Other studies like HCSC, ASC, EuPEO or ZI:EL all ended with policy recommendations regarding physical activity (e.g., Naul & Dreiskaemper, 2012; Naul, Staufenbiel, Dreiskaemper & Schöckel, 2015).

Note

Guideline 40 – The dissemination and implementation of these EU Physical Activity Guidelines at EU level should be supported through an EU HEPA Network based on the existing European network for the promotion of health-enhancing physical activity (HEPA Europe).

As the WGI stands for several projects that aim to implement the EU Physical Activity Guidelines, it managed to connect with other partners who share the same aims and values during the last decades. Collaborations with international federations and associations like CEREPS (*Conseil Européen des Recherches en Éducation Physique et Sportive*), EUPEA (European Physical Education Association), ICSSPE (International Council for Sport Science and Physical Education), ISCA (International Sports and Culture Association), or ENGSO (European Non-Governmental Sports Organisation) for youth have been established for years. Together with these co-workers and other partners like national and international universities and projects targeting the EU Physical Activity Guidelines are tackled regularly in applications for funding, preparatory actions, publications or international projects. Besides, several sport federations in Germany are supporting the WGI so that a reciprocal relationship in promoting physical activity guidelines exist.

12.3 Summary, discussion and perspectives

Within the last 25 years, the WGI has developed to a benchmark for physical activity research in Germany and Europe. Mainly based on the work of Roland and his collaborators, a tremendous number of research projects and publications were realized, congresses were held, expert meetings with policy and communities were organized and several best-practice and recommendation letters were created to assist the society to meet the HEPA guidelines of the EU.

These developments were strongly connected to a lot of motivation and effort that was put into the projects. With the move from Essen to Münster in the last years, the necessity of a reorganization of the WGI became obvious due to the number of projects that Roland has initiated during the last years. Therefore, the new structure of the WGI consisting of four research domains seemed most suitable to tackle the challenges and tasks that emerge in these fields of research. A growing team of researchers is currently working on central research projects on health and physical activity, children and youth sport, grassroot and elite sports, Olympic education and physical education. So, we are looking forward to further projects, applications and publications about HEPA and other related topics in the following years. The fundament for a successful continuation and extension of the work of the WGI in the domain of PA research has – without any doubt – been set by the work of Prof. Dr. Dr. h.c. Roland Naul.

References

Dreiskaemper, D., & Naul, R. (2013). Was hängt mit motorischen Leistungen und BMI im Projekt "Gesunde Kinder in gesunden Kommunen (gkgk)" zusammen? In F. Mess F., M. Gruber M., & A. Woll A. (Eds.),: *Sportwissenschaft grenzenlos?!* , (S. 240). Hamburg: Czwalina.

Dreiskämper, D., & Naul, R. (2011). HCSC – The EU-based preparatory work project (2009–2011): Physical and motor development of young European children compared to normal weight and obese parents. In *7. Jahrestagung der europäischen HEPA vom 10. bis 12.10.2011*, Amsterdam (Niederlande).

Dreiskämper, D., & Naul, R. (2014). Gesunde Kinder in gesunden Kommunen. *Betrifft: Sport*, 36(6), 10–16.

Dreiskämper, D., Utesch, T., Henning, L., Ferrari, N., Graf, C., Tietjens, M., & Naul, R. (2018). Motorische Leistungsfähigkeit und physisches Selbstkonzept als Entwicklungsressourcen im Kindesalter von der Kita bis zur Grundschule. *Bundesgesundheitsblatt*, 12.

Ehlen, S, Dreiskämper, D, Utesch, T, & Rehaag, R. (2018). Integrierte Gesamtansätze kommunaler Gesundheitsförderung für Kinder. Untersuchungsdesign und interdisziplinäre Methodenintegration des Forschungsverbunds "Gesund Aufwachsen". *Prävention und Gesundheitsförderung*, 13(3), 254–260. doi:10.1007/s11553-018-0640-7

EU Working Group. (2008). EU Physical Activity Guidelines recommended policy actions in support of health-enhancing physical activity. Approved by the EU Working Group "Sport and Health".

Forschungsgruppe SpOGATA. (2012). Abschlussbericht des Forschungsprojektes "Evaluation des BeSS-Angebotes an offenen Ganztagsschulen im Primarbereich in seinen Auswirkungen auf die Angebote und Struktur von Sportvereinen, Koordinierungsstellen und die Ganztagsförderung des Landessportbundes NRW in Nordrhein-Westfalen".

Forschungsgruppe SpOGATA. (Ed.). (2015). *Evaluation der Bewegungs-, Spiel-und Sportangebote an Ganztagsschulen in Nordrhein-Westfalen* (Vol. 15). Aachen: Meyer & Meyer Verlag.

Heemsoth, S., Dreiskämper, D., Tietjens, M., & Naul, R. (2012). 10 Welche Rahmenbedingungen fördern und behindern eine gesundheitlich-motorische Entwicklung von Grundschulkindern im Ganztag?

Magro, S., Utesch, T., Dreiskaemper, D., & Wagner, J. (2018). Self-esteem development in middle childhood: Support for sociometer theory. *International Journal of Behavioral Development.* doi:10.1177/0165025418802462

Naul, R. (2014). The European Union and Olympism: "coubertin plan" and youth Olympics. *Journal of Olympic History*, 22(2), 53–59.

Naul, R., Schmelt, D., Dreiskämper, D., Hoffmann, D., & l'Hoir, M. (2012). "Healthy children in sound communities" (HCSC/ gkgk) – A Dutch-German community-based network project to counteract obesity and physical inactivity. *Family Practice*, 29(2), 110.

Heemsoth, S., Tietjens, M., Dreiskämper, D. & Naul, R. (2015). Welche Rahmenbedingungen fördern und behindern eine gesundheitlich-motorische Entwicklung von Grundschulkindern im Ganztag? *In Forschungsgruppe SPOGATA (Hrsg.), Evaluation der Bewegungs-, Spiel- und Sportangebote an Ganztagsschulen in Nordrhein-Westfalen*, (pp. 201-214). Aachen: Meyer & Meyer.

Naul, R., Schöckel, Y., & Dreiskämper, D. (2016). Qualitätsentwicklung ZI:EL – Bericht zur Gesamtevaluation und wissenschaftlichen Begleitung des Projekts "Zukunftsinvestition: Entwicklung jungen Engagements im Sport." Essen/Münster: WGI.

Naul, R., Utesch, T., & Dreiskämper, D. (2018). Monitoring motor development: Measuring motor competence via an online database. In C. Scheuer, A. Bund, & M. Holzweg (Eds.), *Changes in childhood and adolescence. Current challenges for physical education* (p. 76–80). Berlin: Logos.

Naul, R., Utesch, T., & Niehues, D. (2018, in Druck). Linking physical education with local sport organizations: Implementation of an "active school community". *European Journal of Physical Education & Sport Science*, *4*(11), 1–27. doi:10.5281/zenodo.1411519

Shen, J., Naul, R., Dreiskaemper, D., & Heemsoth, S. (2015). Investigation research on health promotion and physical activity intervention of primary school students – Take two

Shanghai District Schools as examples. *Sport (Chinese)*, 120(16), 54–56.

Shen, J., Naul, R., You, Y., & Dreiskämper, D. (2014). Research on Sino-European Health Project "Healthy Children in Sound Community." *Sport Science Research (China), 35*(6), 90–94.

Utesch, T., Dreiskämper, D., Naul, R., & Geukes, K. (2018). Understanding physical (in-) activity, overweight, and obesity in childhood: Effects of congruence between physical self-concept and motor competence. *Scientific Reports*, 8, 5908. doi:10.1038/s41598-018-24139-y

Utesch, T., Dreiskämper, D., Strauss, B., & Naul, R. (2018). The development of the physical fitness construct across childhood. *Scandinavian Journal of Medicine & Science in Sport*, 28(1), 212–219. doi:10.1111/sms.12889

Utesch, T., & Naul, R. (2015). Die gkgk-Online-Datenbank zum Monitoring der gesundheitlich-motorischen Entwicklung von Kindern und Jugendlichen. *WGI Newsletter* Nr. 9. Münster: WGI.

13 Erasmus+-funded HEPA: Promoting projects in Europe

Project activities of the European Physical Education Association

Claude Scheuer, Rose-Marie Repond & Martin Holzweg

13.1 Introduction

The European Physical Education Association (EUPEA) was founded in 1991 in Brussels and brings together as the umbrella organization of the national Physical Education Associations in Europe members from more than 30 European countries Europe. The mission of EUPEA is focusing on the promotion of and the advocacy for physical education in Europe, described in the Declaration of Madrid from 1991, amended in Brussels 2011, with the main slogan 'No Education without Physical Education'.

From 2014 to 2015, EUPEA started its activities in European projects in two Preparatory Actions for European Partnerships on Sport:

- The *Pro Safe Sport project (PSS)* coordinated by the Enlarged Partial Agreement for Sport (EPAS) of the Council of Europe, elaborating guidance, recommendations and training to sport administrators and coaches on physical, psychological and social well-being in sport.

- *Recall: Games of the Past – Sports for Today*, a project coordinated by TAFISA and supported by 12 partner organizations coming from all over Europe, aiming to reintroduce traditional sports and games into the daily lives of young people as a means to tackle social challenges of today.

Since the implementation of the new Erasmus+ programme by the European Commission in 2014, one of EUPEA's core strategies to implement its mission is to participate in European projects based on collaborative partnerships.

13.2 About ERASMUS+

The new Erasmus+ programme of the European Commission aims to support actions in the fields of education, training, youth and sport for the period 2014–2020. Erasmus+ replaces seven programmes bringing together

- The Lifelong Learning Programme (Erasmus, Leonardo da Vinci, Comenius, Grundtvig and Jean Monnet)
- The Youth in Action Programme
- Five international cooperation programmes (Erasmus Mundus, Tempus, Alfa, Edulink, the programme for cooperation with industrialised countries)
- The new sport action

The programme supports organizations to work in transnational partnership and to share innovative practices in the fields of education, training and youth. Erasmus+ also includes a strong international dimension (i.e., cooperation with partner Countries), which opens the programme to institutional cooperation.

The Erasmus+ Sport Programme specifically supports collaborative partnerships, small collaborative partnerships and non-profit sport events. These actions are expected to result in the development, transfer and/or implementation of innovative practices not only at European and national, but also at regional and local levels and to contribute to the development of a

European dimension of sport. In addition, these actions are also supporting the implementation of the European Commission's initiatives to promote sport and physical activities in the EU.

With a budget of over €2 billion in its initial year, Erasmus+ has already offered more than one million people the opportunity to take part in 18,000 funded projects. More flexible opportunities for collaboration across sectors are also helping Europe's education, training, youth and sport systems to try out innovative practices and contribute to reform and modernisation.

13.3 The ASC project

The Active School Communities (ASC) project, implemented in 2016 and 2017, aimed to equip community sport organizations with an adaptable toolkit they can use to collaborate more effectively with schools on physical activity initiatives for school students. The toolkit also assists them in establishing contact with and advocating towards governments to drive the impact of their initiatives. The knowledge base used to develop the toolkit was drawn from an experienced group of partners from research institutions, the education sector and non-profit sport organizations from nine countries in west, east, north and south Europe. Amongst these partners, the Willibald Gebhardt Institute under the leadership of Roland Naul contributed to the success of the project.

The project's emphasis on strengthening the links between community sport organizations and schools offered a low-cost solution to improve physical activity and physical education in schools, as well as opportunities for cooperation by forging partnerships between stakeholders who are already working at the community level, but often separately, to encourage children to move more.

The project contributed to address the EU's ambitions to enhance collaboration between the education and grassroots sport sectors. The overall objective of the ASC project was to build the capacity of community sport organizations to harness potential partnerships in the school setting for the promotion of sport and physical activity for healthy lifestyles and better learning outcomes.

As research shows, too many people across Europe are not physically active enough. This has a serious impact on general health, the frequency of preventable diseases and the number of premature deaths. The World Health Organisation and the European Commission recommend that school-aged young people should participate in 60 minutes or more of moderate to vigorous physical activity daily, in forms that are developmentally appropriate, enjoyable and involve a variety of activities.

Sports and physical activity organizations and schools are well placed to address the decline in physical activity and to provide an appropriate offer of physical activity, particularly if they work in partnerships. The ASC toolkit was designed to improve the effectiveness of these partnerships.

The toolkit includes

- advocacy documents for sports organizations to share with local and national policymakers, young people, parents and schools in order to explain the benefits of increased physical activity and the contribution of the ASC programme in particular;
- a self-assessment tool for sports organizations to evaluate their commitment and capacity to work in school settings to increase physical activity in all young people;
- an action-planning tool to enable sports organizations and schools to communicate and work effectively in partnership to increase physical activity for healthier lifestyles and better learning outcomes in their students.

The project, jointly funded by the European Commission's Erasmus+ programme and 12 European and national partners, tested the adaptability and viability of the toolkit in different national settings by trialling a pilot version in six European countries and arrived at a resource that was rolled out Europe-wide in 2017.

The partnership was designed to foster interaction between community sport, education and government sectors across Europe in order to establish more inclusive and cohesive networks of community stakeholders in the promotion of health-enhancing physical activity (HEPA).

13.4 The ActiveVoice project

The overall objective of this project implemented between January 2016 and December 2018 was to build capacity in civil society organizations to engage in active advocacy for the implementation of the EU Physical Activity Guidelines. The European partners in this project were mainly umbrella organizations with experience in advocacy on the EU level. They engage, in turn, with their member associations (national project partners and beyond) to deliver capacity building, alliance creation and specific advocacy interventions on national and local levels.

The limited progress made since 2008 to implement the EU's Physical Activity Guidelines, as mapped so far by the former Expert Group on Sport, Health and Participation, and most recently by the National Focal Points, shows several gaps in policymaking and structures in the field of HEPA in the EU. Over these years, HEPA promotion has either not been treated as a major issue by national and local governments, or it has been addressed in a narrow policy context (i.e., by one ministry only) or in a fragmented way (multiple, non- coordinated policy interventions).

The ActiveVoice project partners believe that guidelines, recommendations and the corresponding soft monitoring of them are not likely sufficient in themselves to ensure an effective and consistent implementation across Europe. This approach, which may be called 'top-down', is important for the policy context and the awareness and the benchmarking implied can have some agenda-setting effects. However, without a consistent push from other stakeholders beyond the EU and the governments themselves, that is, a 'bottom- up' policy implementation approach, the progress is likely to remain limited.

To achieve better progress, civil society organizations need to play a more significant role in pushing the agenda forward, in particular by advocating for better implementation of the guidelines to national and local governments. However, many organizations promoting grassroots sport and physical activity in the EU member states do not have the capacity to deliver advocacy, and many simply do not know where to start.

13.4.1 Giving grassroots sport organizations a more 'ActiveVoice' in physical activity policy

The ActiveVoice project aimed at developing relevant tools and resources that will in the future help enhance these organizations' influence on physical activity policy and make their advocacy efforts more effective overall. The project partners were European umbrella organizations with experience in advocacy at the EU level. During the project, they engaged with their member associations (the national project partners and beyond) to deliver capacity building and facilitate alliance creation and specific advocacy interventions on national and local levels.

13.4.2 The ActiveVoice project partners

International partners:
International Sport and Culture Association (ISCA), EPODE International Network (EIN), European Cyclists' Federation (ECF), European Healthy Stadia Network, European Physical Education Association (EUPEA),

National partners:
International Association of Sport and Leisure Infrastructure (Slovenia), UFOLEP (France), V4Sport (Poland), BG Be Active (Bulgaria), UISP (Italy) and Greenways SCE (Greece).

13.4.3 Key topics

The ActiveVoice project partners selected topics in which they have expertise and wished to initiate or facilitate advocacy. They worked together with members and partners they had chosen to support through the project in the following fields:

- physical activity and health
- cycling, transportation and active travel

- sport facility management and urban planning
- education and active schools
- obesity and behaviour change

13.4.4 Project objectives

The overall project objective was to build capacity in civil society organizations to engage in active, cross-sector advocacy for the implementation of the EU Physical Activity Guidelines.

The specific project objectives and expected deliverables were the following:

- To build the knowledge base for active advocacy for the implementation of the EU Physical Activity Guidelines by
  - determining the actual level of progress in implementing the EU Physical Activity Guidelines in five selected EU countries in cooperation with the national HEPA focal points;
  - compiling good practices in the field of advocacy for physical activity amongst participating organizations and beyond through desk research from past projects, other alliances and relevant literature;
  - commissioning pilot reports in Poland, Bulgaria and France, establishing an evidence base for active advocacy for HEPA promotion. These pilot researches allowed international comparison and are adaptable to different settings.
- To build capacities in participating organizations to engage in active advocacy by
  - developing relevant tools and resources to facilitate the advocacy work of all involved organizations;
  - conducting knowledge-sharing and training seminars/workshops.
- To test the validity and quality of the proposed tools and capacities by
  - establishing cross-sector HEPA advocacy alliances at the national and local level;
  - delivery of advocacy interventions.
- Widening the impact of the project by
  - developing a webinar series on HEPA policy advocacy;
  - dissemination and exploitation of the project results.

13.5 The EuPEO project

The European Physical Education Observatory (EuPEO) project, coordinated by the Faculdade de Motricidade Humana of the University of Lisbon and running from January 2018 until December 2020, will inform the platform of a future European Physical Education Observatory, bridging the gaps between guidelines, concepts for monitoring and evaluation, sport, health and education sectors, and between different sets of indicators, to foster quality physical education (PE), school sport (SS) and other forms of promotion of School Physical Activity in Europe.

To build this monitoring solution based on the collaborative work between professional and scientific institutions, 12 partners have developed together the EuPEO project proposal, including EUPEA, EUPEA member PE teacher associations and universities in eight European countries. Amongst these partners, the Willibald Gebhardt Institute, under the leadership of Roland Naul, will contribute to the implementation of the project in Germany in close cooperation with the German PE teacher association (Deutscher Sportlehrerverband, DSLV).

The EuPEO project aims to convert former monitoring and evaluation experiences of PE, HEPA, the Public Health and Education sectors into a comprehensive and applicable monitoring system by developing a EuPEO web page, a manual for external monitoring (MEA) and a toolkit to prepare and provide internal self-monitoring (TIM) of quality PE and SS. During a three-year study, the partners will be involved in the construction and validation of these instruments, as well the dissemination and exploitation of the results.

The MEA will be developed as an open-access methodological e-book to be integrated in the platform supported by the EuPEO web page to guide the data-collection process feeding the observatory database. Moreover, the MEA will be important for advocacy and for capacitating Europe to monitor the HEPA recommendations. The TIM will capacitate schools to monitor HEPA recommendations directly.

Eight national multiplier sport events will disseminate this platform in each partner country. One European multiplier sport event, the EUPEA Forum 2020, will discuss the platform with other international institutions to become part of the project. Furthermore, each country will be able to identify good examples, disseminate them and motivate other schools to implement similar conditions.

So far, the EuPEO project partners have finished the data collection on indicators of quality PE, SS and school-based physical activity for the validation of a European Country Questionnaire (ECQ), a European School Questionnaire (ESQ) and National External Learning Assessment Systems (NELAS) in PE.

13.5.1 Rationale

The role of PE has been highlighted as the only curriculum subject whose focus combines the body and physical competence with values-based learning and communication, provides a learning gateway to grow the skills required for success in the twenty-first century, as citizenship promotion and health promotion through a lifelong participation in physical activity, social inclusion through the challenging of stigma and stereotypes and academic achievement by improving children's attention span, enhancing their cognitive control and speeding up their cognitive processing.

European policies and legislation, research and high-profile reports in the field of sport and education identified the necessity of regular monitoring of PE, SS and PE teacher training, across the 28 European Commission member states. In order to harmonise the conditions of its educational offer for the promotion of an active and healthy citizenship, evidence is necessary to design and to develop a manual and a toolkit to monitor, with an extended frame of indicators, the quality assurance of PE and HEPA in curricular and in extra-curricular time in school settings, especially considering the need for inclusion of disadvantaged populations.

13.6 The ESSD project

13.6.1 History

The European School Sport Day (ESSD) has empowered nearly 2.5 million European students to be active in 30 countries so far. The Hungarian School Sport Federation started to pilot ESSD in Europe in 2015 in an Erasmus+-funded project with the aim of connecting Europe through the organization of a European School Sport Day, supported by the development of a methodological toolkit to provide scientific background for the respective school sport events. This initiative quickly expanded across Europe and by linking it with the European Week of Sport, it was possible to get more and more schools on the move. As a partner in the ESSD consortium, EUPEA involves its national members in the coordination of ESSD activities in their respective countries, as for example, the Willibald Gebhardt Institute with Roland Naul in parts of North-Rhine Westphalia in Germany.

13.6.2 Objectives for participating schools

- Get all students active for at least 120 minutes on the ESSD
- Raise the profile of PE and sports in schools
- Promote health and well-being for lifelong learning
- Encourage social inclusion and develop social competences amongst students
- Connect with other European countries

13.7 The PASS project

The overall objective of the Physical Activity Serving Society (PASS) project implemented from 2015 to 2017 was to tackle the physical inactivity crisis in Europe, which has been documented so far essentially through the 2008 EU Physical Activity Guidelines at the institutional level. The project aimed at gathering and analyzing European data on the scale, costs and consequences of the physical activity crisis in Europe, with the view to complementing existing European data on the state of physical activity in Europe and to identify and promote sectorial good practices while adopting an innovative cross-sectorial, holistic and transversal approach.

13.7.1 Objectives and missions

The main goal of the PASS project was to propose political and strategic actions in order to promote physical activity in Europe. It was inspired by the Designed to Move campaign, already launched on several other continents.

More specifically, the starting point of the reflection was an alarming situation: levels of physical inactivity have reached an unprecedented peak in Europe. Indeed, according to the latest Eurobarometer survey on Sport and Physical Activity, 42 per cent of EU citizens do not engage in physical activity at all.

Consequently, the project aimed at

- documenting the scale, costs and consequences of the physical activity crisis in Europe;
- raising awareness of the physical inactivity crisis to decision makers and the need to make HEPA more of a transversal political priority;
- engaging change-makers and sport prescribers in critical sectors to show how HEPA can align with their objectives.

13.7.2 Methodology

The PASS project was led according to a three-step methodology:

- First, to gather and analyze European data on the scale, costs and consequences of the physical activity crisis in Europe;
- then, to use this research to raise awareness of the physical activity crisis to decision makers – and the need to make HEPA more of a political priority;
- and finally, to engage change-makers in critical sectors to show how HEPA can benefit their work and help solve the challenges faced in these areas.

13.8 The SHAPE project

The Shaping the Principles of Physical Education (SHAPE) project was implemented under the leadership of the Hungarian School Sport Federation from 2015 to 2016. The official name of the project was 'Shaping the principles and development areas for health-oriented physical education based upon the comparative analysis of the core curricula in some EU member states in order to establish a common European framework', SHAPE for short. This is highly indicative of the primary objective of cooperation, namely that EU member states subject their documents governing PE to comparative analysis and use this analysis to establish a common framework providing an objective benchmark for implementing quality PE at the member-state level. Besides Hungarian School Sport Federation (HSSF), the project partners were the Ministry of Education and Religious Affairs of Greece, the Johann Wolfgang von Goethe University, Department of Sports Medicine, from Germany; the Kaunas University of Technology, from Lithuania; as well as the University Ljubljana, from Slovenia. The EUPEA was involved actively in the project as a contributing expert.

The essential goal of the SHAPE project was to compare regulatory documents and systems of school PE in the affected member countries, and – based on this analysis – to give recommendations for concrete interventions on developing health-oriented quality

PE by establishing a European Framework of Quality Physical Education. Indirectly the project aimed to incite a discussion on establishing a new Lifelong Learning (LLL) Key Competence area related to health-oriented life management within the European Qualification Framework.

As a result of the project through individual developmental work and several common workshops, the members of the consortium produced analyses and policy recommendations in the field of school PE in the following areas:

- National qualification frameworks
- Role of the subject PE in the public education system
- LLL Key Competences
- National core curricula and framework curricula

The results of the expert consortium's developmental work were disseminated in HSSF's biannual international conference on quality PE (named HIPE for HEPA) in November 2016.

13.8.1 Outputs of the project

The main deliverables and outcomes of the project were the following:

- a meta-analysis on the national qualification frameworks, regulatory documents of the public education systems (PE-related contents), LLL key competences;
- a meta-analysis on the national core curricula;
- national recommendations (where relevant) on the respective public education systems;
- a recommendation on the European Framework for Quality Physical Education based on the European Qualification Framework;
- a recommendation for a new LLL competency: Health-Conscious Future-Oriented Life Management Key Competence;
- an e-book summarising the findings of the two-year project.

As a result, the project developed the European Framework for Quality Physical Education, which presents the necessity of focusing on the quality of PE, including learning outcomes specified for students in four levels. With respect to PE, this framework designates five main development (learning) areas, as well as associated knowledge, skill and competency contents. The latter are pivotal for secondary-school leavers in order to lead a health-conscious lifestyle that includes regular physical activity.

In the frame of the project was also developed a proposal for a *Health-Conscious Future Oriented Life Management Key Competence*, which could lay the foundation for the establishment of a new key competency area in European school systems, as well as in the LLL key competences' benchmark, determining competency components required for a health-conscious, physically active future-oriented lifestyle.

This project was recognised by the European Commission as a 'success story'.

13.9 New projects started in 2018

The project PASS closed at the end of 2017 under the lead of the French think tank Sport & Citizenship but will have a follow-up with the three-year-project *PACTE (Promoting Active Cities throughout Europe)*.

In another two-year-project under the lead of the University of Potsdam with the name *BMC-EU (Basic Motor Competencies in Europe – Assessment and Promotion)*, EUPEA will contribute with its expertise to the implementation of competence-oriented motor tests and a toolkit for teacher training in 12 European countries.

A further two-year project is under the lead of the HSSF (Hungarian School Sport Federation). EUPEA will contribute in the project *EU-PALS (European Physical Activity Label for Schools)* to the development of a European label for active schools.

In another application, the HSSF has been successful with an application for a follow-up project based on the two previous projects developing an *ESSD*, with the objective of implementing this initiative in more European countries. The EUPEA is contributing to the success of this project as a partner.

Furthermore, the EUPEA is a key partner in the small collaborative project *PHYLIT (Physical Literacy)* under the lead of Youth Sport Trust International. The aim of this project is to introduce physical literacy as key competency of the European Council Framework for Key Competences for Lifelong Learning.

Finally, the EUPEA will contribute to the development of a free open educational resource (web application) for teachers to access information, pedagogical tools and resources to include all children in primary PE lessons in the project *DIPPE (Disentangling Inclusion in Primary Physical Education)* coordinated by the University of Luxembourg. This is the first project for EUPEA in the Key Action 2 of the Erasmus+ programme on Strategic partnerships in School Education.

13.10 Conclusion and outlook

In the past four years, EUPEA has participated in many European projects. Starting these activities in the Preparatory Actions for European Partnerships on Sport by being asked to join the respective project consortiums by other associations, EUPEA was meanwhile able to develop successfully its own project applications (EuPEO and DIPPE), together with its national members. In the future, EUPEA will continuously develop its strategy when it comes to a participation in European projects. As EUPEA is an association focusing on PE as a learning subject in educational systems and as the challenges of EUPEA's members are mainly situated in the implementation of quality PE, the Key Action 2 for strategic partnership in different fields of education of the Erasmus+ programme will be the target area for EUPEA in the future.

Project references

Active School Communities (ASC): activeschoolcommunities.eu

ActiveVoice: www.activevoice.eu

European Physical Education Observatory (EuPEO): www.eupeo.eu

European School Sport Day (ESSD): www.essd.eu

Promoting Active Cities Throughout Europe (PACTE): www.sportetcitoyennete.com/en/europe/pacte-project

Physical Activity Serving Society (PASS): sportetcitoyennete.com/pass

Pro Safe Sport (PSS): pjp-eu.coe.int/en/web/pss

Recall: Games of the Past – Sports for Today: recallgames.com

Shaping the Principles of Physical Education (SHAPE): www.mdsz.hu/en/shape

14 HEPA as 'le geste sportif'? Structures versus activities in public policies for the promotion of sport and health-enhancing physical activity

Jacob Kornbeck[3]

Now there is nothing in this world I abominate worse, than to be interrupted in a story. . .

Laurence Sterne[4]

If and when public authorities commit themselves to support health-enhancing physical activity (HEPA), including but not limited to sport, they will face a number of conceptual issues: what should they fund, and/or who should they fund? Should they earmark public funds for specific organizations or for specific activities? If public authorities are committed to HEPA and 'sport for all', how can they support the relevant activities and/or structures while not disengaging from a more traditional national sport policy track, geared towards nation-branding, national pride, harvesting medals and hosting events?

---

3 *Previously policy officer in the European Commission, Sport Unit (2001–14). Currently policy officer in the European Commission, Youth Unit.* Disclaimer: *Remarks made strictly in a private capacity. No official EU positions expressed.* Acknowldgement: The author would like to thank Prof. Bjarne Ibsen (Odense, Denmark) for his comments on previous versions of this text.

4 *Laurence Sterne, The Life and Opinions of Tristram Shandy, Gentleman (1759 et seq.), Chapter 3.LXXXIV.*

This chapter will argue in favour of a conceptual framework relying on the notion 'le geste sportif' (approx. 'the sporting act', 'the sporting movement') derived from EU sport-relevant case law (section 14.2). Having isolated this concept for heuristic purposes, the chapter will progress to employ it as an analytical tool for a reassessment of the HEPA policy choice of structures versus activities, by first examining a range of key recommendations enshrined in the 2008 EU Physical Activity Guidelines, to which Prof. Roland Naul (to whom this book has been dedicated) made a substantial contribution (section 14.1), and secondly by looking into a case study consisting of the Danish sports lotteries legal framework (section 14.2). The findings will be discussed with reference to policies in some other EU countries (section 14.3), and the chapter will be rounded off by a conclusion (section 14.4), pointing to an increased relevance, in sport policy contexts marked by austerity policies, of the matters discussed in this chapter.

14.1 HEPA as 'le geste sportif'?

One way of framing a public policy pledge to support HEPA would be – in line with established national practice in many national contexts – to reserve funding for specific organizations with a sports delivery remit. In this case public funding could be made available with or without 'strings attached': that the latter option is not only an 'academic' but also a well-probed practical one can be seen from the Danish case study (section 14.2). Allowing for some variation, therefore, this bunch of options can be phrased as 'funding targeting structures'. But if the focus of public funding is not on structures (which seems straightforward and easy from a bureaucratic logic, besides promising attractive political alliances with sports organizations, again triggering political capital), what would the alternative approach consist in, and what should it be called? In this chapter it will be referred to as 'funding targeting activities', because under this philosophy, activities deemed worthy of public funding will be the focus, *whether or not* they are offered by established sports organizations. On the one hand such organizations are very welcome to apply for funding, while their great potential for outreach predicts that they can be strong on delivery. On the other hand, other civil society organizations will also be eligible, within the frame of open calls for proposals, if their activities and their project proposals meet the published criteria.

In legal terms, the Court of First Instance (CFI)[5] of the EU may unwittingly have paved the way in its *Meca-Medina* decision of 30 September 2004,[6] which is otherwise more widely known for having addressed issues related to the applicability of competition law (antitrust) rules to anti-doping rules, and especially for having been subsequently quashed by the court on 18 July 2006[7] (for details, see Kornbeck, 2017b). What the CFI also did, however, was to enunciate a theory of 'sport' in its purest, un-commercial form. The enormous significance of *Meca-Medina* lies in the fact that the court arrested such speculations, which had been nurtured by ambiguities in the sport-related case law ever since its first occurrence in the court's *Walrave and Koch* judgment (1974). It was only natural and legitimate that sports governing bodies (SGBs) expected some aspects of their regulations to be *per se* exempt from EC/EU free movement rules and later, as in *Meca-Medina*, also from antitrust rules. In quashing the CFI's decision, the court made it clear that sporting rules may at any time be considered in possible breach of antitrust law: a case-by-case approach is needed, each case being evaluated on the basis of its desired outcomes and real-term economic effects. The CFI referred to 'sport' in its English-language version, but to 'das Sporttreiben' in German and to 'le geste sportif' in French. The court, by contrast, referred to 'sport as such', 'Sport als solcher' and indeed 'le sport en tant que tel', respectively. But despite having been overruled by the court as regards the applicability of EU law, the CFI had a lucky hand in picking the concept 'le geste sportif', for in French (in this case the language 'of procedure' elicited by the claimants, and thus the 'original version' best placed to provide the basis for a semantic interpretation of the judgment), 'geste'

5 *While today, cases are adjudicated first by the General Court and then by the Court of Justice of the European Union, at that time the term was Court of First Instance. Before 1993 there was only one instance.*

6 *Judgment of the Court of First Instance (Fourth Chamber) of 30 September 2004. David Meca-Medina and Igor Majcen v Commission of the European Communities. Case T-313/02. ECR 2004 II-03291. ECLI:EU:T:2004:282.*

7 *Judgment of the Court (Third Chamber) of 18 July 2006. David Meca-Medina and Igor Majcen v Commission of the European Communities. Case C-519/04 P. ECR 2006 I-06991. ECLI:EU:C:2006:492.*

refers to physical movements *as well as* to any other concrete actions; they may or may not be physical, yet actions they are: they *happen.* It has been argued elsewhere that 'le geste sportif' could be an interesting alternative to the 'special characteristics of sport' referred to in Article 165 TFEU, which have proven difficult to apply in practice. And while such distinctions may still seem dubious in relation to free movement and antitrust rules, they may be very useful in defining priorities for HEPA funding.

14.2 Structures versus activities in public policies for the promotion of sport and HEPA

14.2.1 The 2008 EU Physical Activity Guidelines

Public sport policy grants may take the form of 'basic grants' *or* 'performance grants' (Ibsen, 2014), respectively. This distinction is of particular importance given the rising financial burden born by public authorities who have chosen to participate in the race for hosting international events (Kornbeck, 2015a). A powerful example was provided recently in the shape of the referendum held in Hamburg, on 29 November 2015, the outcome of which was that the public authorities had to withdraw their financial guarantees underpinning the bid which, consequently, had to be retracted. In contrast with the traditional sport policy pursued by many countries, the HEPA concept had initially been developed within a World Health Organisation (WHO) framework but was, in the first two decades of the new millennium, to become instrumental in shaping a novel EU approach to sport policy. This approach is apt to show member states an alternative way, as opposed to a traditional sports policy aimed at harvesting more medals at international competitions, at hosting as many of these competitions as possible and, in so doing, at aligning public policies as closely as possible with the preferences and expectations of sports organizations, including at international and national level. This alternative sport policy ideal is neatly encapsulated in the EU Physical Activity Guidelines (European Commission, 2008), which include the following policy recommendations:

Guideline 6 – 'When public authorities (national, regional, local) support sport through public budgets, particular attention should be given to projects and organisations which allow a maximum of people to engage in physical activity, regardless of their level of performance ("sport for all", recreational sport).'
European Commission, 2008, p. 16

Guideline 7 – 'When public authorities (national, regional, local) support sport through public budgets, appropriate management and evaluation mechanisms should be in place to ensure a follow-up that is in line with the objective of promoting "sport for all."' *ibid., p. 16*

Guideline 8 – 'When public grants are made available for activities with physical activity content, the eligibility and allocation criteria should be based on the activities foreseen, and on the general activities of the organisations applying for funding. A specific legal status, organisational history or membership in larger federative structures should not be considered as pre-qualifying. Funding should be directed toward 'sport for all' activities, bearing in mind that organisations with an elite sport component may also make a meaningful contribution to the "sport for all" agenda. Public and private actors should be able to compete for funding on an equal footing.' *ibid., 2008, pp. 16–17*

Guideline 9 – 'Sport policy should be evidence-based and public funding for sport science should encourage research that seeks to uncover new knowledge about activities that allow the population at large to be physically active.' *ibid., p. 17*

EU action in the field of sport is considerably curbed by the EU's mandate being limited by the treaty's principle of 'conferral' (Article 5 TFEU) (Treaty on the Functioning of the European Union) (whereby the Union can only take legislative action when member states have specifically empowered it to do so via an explicit treaty provision). The EU has nevertheless been able, for quite some time already, to use soft law instruments to push for better sport governance (Kornbeck, 2015a), although admittedly, the (unstated, subdued or implied) threat of hard law action may at times have acted as a catalyzer vis-à-vis sports organizations (Kornbeck, 2006). EU market legislation may have paved the way for a 'Normative Market Europe' (Geeraert & Drieskens, 2017) although, under the pre-Lisbon Treaty framework, the absence of an explicit reference to sport in the treaty made this exercise more cumbersome. The European Year of Education through Sport (EYES 2004) had to be grounded in the EU education competence (ex-Article 149 TEC) to make the initiative lawful at all. Since the entry into force of the Lisbon Treaty, however, the enhanced Article 165 TFEU (Treaty on the Functioning of the European Union) has offered an opportunity to further apply this approach: 'Lisbonisation without Regulation' (Kornbeck, 2015a). A parallel example would be the case of EU Guidelines on so-called dual careers, defining good practice regarding athletes' access to non-sporting qualifications as a means to reduce the risks incurred when retiring from their active sporting careers (Kornbeck, 2017a). Yet to play this new-found role, the EU needed a sport policy doctrine.

After the European Commission had commissioned an external study aimed at identifying trends in young people's physical (in)activity, fitness and (over)weight, and the ensuing policy implications and challenges (Brettschneider & Naul, 2005), a range of HEPA-style projects started to emerge, in various local contexts (Brettschneider, Hoffmann, Naul & Steinzen, 2009), following the recommendations of that study (Brettschneider & Naul, 2005). Together with other strands of scientific knowledge and practical know-how, including from the implementation and evaluation of EYES 2004 (Eureval-C3E, 2005), an understanding started to emerge regarding how the HEPA concept could be applied within the EU. In relation to the fundamental question of whether sport policy grants should rather support 'basic grants' or rather 'performance grants' (Ibsen, 2014), the EU chose the latter, even if critical observers may take the view that many national grants continue following traditional patterns rather than the agreed EU-wide HEPA approach. Following an invitation from member states' sport ministers, an expert group proposed a set of HEPA recommendations, termed 'EU Physical Activity Guidelines' (European Commission, 2008), which have since then formed the standard reference framework for comparing sport policy measures in view of their HEPA-friendliness.

While the actual influence exercised on national policies may still be questionable, apart from the possibility of sharing information and good practice in an area without a hard law competence (Kornbeck, 2015b), this HEPA doctrine has durably and visibly pervaded the EU's own grants policies, which *do not* earmark funding for specific organizations, but rather allows *all* civil society organizations and public authorities to apply, requesting them to demonstrate the HEPA-relevance of the proposed activities (European Commission, 2017). Recent decisions of the commission and the EU courts[8] in the field of state aid, where aid is often deemed compatible with the single market when the activities or structures aided are open to the general public, while financial aids to ailing footballs clubs usually are not deemed compatible (García, Vermeersch & Weatherill, 2017), seem to confirm this. At this writing (December 2017), a glance at the list of sport-related competition law cases decided by the commission (state aid, antitrust and merger control)[9] revealed that, during the five years 2015–2017, more state aid cases had been decided favourably[10] than unfavourably.[11] A clear picture emerges from this evidence of decisional practice: aids paid to specific infrastructures are more likely to be compatible with the single market than aids serving as cash injections to ailing (often football) clubs; and the probability that an aid would be compatible and permissible is proportionate to the level of access offered to the general population. These observations seem to confirm the impression of an emerging EU HEPA soft law doctrine and actually one with hard law implications (Kornbeck, 2018). As such, HEPA-based sport policies may be seen as an alternative and a corrective to traditional sports policies.

8 *See, for example, Judgment of the General Court (Eighth Chamber) of 12 May 2016. Hamr Sport a.s. v European Commission. Case T-693/14. Electronic reports. ECLI:EU:T:2016:292. Judgment of the General Court (Sixth Chamber) of 9 June 2016.* Magic Mountain Kletterhallen *GmbH and Others v European Commission. Case T-162/13. Electronic reports. ECLI:EU:T:2016:341.*

9 *See http://ec.europa.eu/competition/sectors/sports/news.html.*

10 *04.07.2016 State aid: Commission clears support measures for certain football clubs in the Netherlands – 09.04.2014 State aid: Commission approves £110 million aid for three sports stadiums in Belfast, Northern Ireland – 18.12.2013 state aid: commission approves French aid for construction and renovation of stadiums for UEFA EURO 2016 Championship – 20.11.2013 state aid: commission clears Belgian support scheme for football stadiums – 15.05.2013 state aid: commission authorises public co-financing of Copenhagen multi-arena in Denmark – 02.05.2013 state aid: commission authorises public co-financing of Uppsala arena in Sweden – 21.03.2012 State aid: Vice President Almunia and UEFA president Platini confirm Financial Fair Play rules in professional football are in line with EU state aid policy – 09.11.2011 state aid: commission clears Hungarian sport support scheme.*

11 *04.07.2016 state aid: commission decides Spanish professional football clubs have to pay back incompatible aid – 01.10.2014 state aid: commission decides Nürburgring racetrack in Germany received incompatible state aid – 19.09.2012 state aid: commission orders recovery of Ð10 million unlawful aid from Slovenian ski maker Elan.*

14.2.2 Case study: The Danish sports lotteries legal framework

Danish legislation on sports betting and the distribution of revenue thus generated originates in rules and institutions stretching back to 1948, representing a compromise between a traditional public policy objective aiming at curbing the gambling activities of the populace, on the one hand, and a wish to let sports organizations and other civil society organizations profit from the revenue which could be generated by allowing such activities regardless on the other hand. Prior to the liberalisation of sports betting which was triggered by the court's *Gambelli*[12] judgment in 2003, the system was operated as a monopoly managed by the sports gambling agency Dansk Tipstjeneste (now Danske Spil). The current legislative framework consists of an act of parliament dated 2010, as amended by a further act of parliament dated 2014 and implemented through a statutory instrument.[13] As defined in the statutory instrument, net revenue generated through authorised gambling operations related to sports competitions, horse and dog races as well as certain lotteries, is to be distributed, after deduction of taxes, fees, prize money, administrative costs, depreciation, provisions, and so on, in accordance with the 2010 and 2014 acts (§ 1 LBK 115[14]). Net revenue is transferred to the ministries of culture (81.24%), education (1.61%), the environment (3.39%), children, gender equality, integration and social affairs (10.42%), health (2.95%) as well as higher education and research (0.39%) (§ 2 LBK 115). Given the rationale of sharing the revenue between different government departments, the very low distribution keys attributed to such

12 *Judgment of the court of 6 November 2003. Criminal proceedings against Piergiorgio Gambelli and Others. Reference for a preliminary ruling: Tribunale di Ascoli Piceno – Italy. Case C-243/01. ECR 2003 I-13031. ECLI:EU:C:2003:597.*

13 *Lov om udlodning af overskud fra lotteri samt heste- og hundevæddemål, LOV nr 696 af 25/06/2010, as amended by Lov om ændring af lov om udlodning af overskud fra lotteri samt heste- og hundevæddemål og lov om fremme af dopingfri idræt, LOV nr 1528 af 27/12/2014, and as implemented through various administrative measures, most recently Bekendtgørelse af lov om udlodning af overskud fra lotteri samt heste- og hundevæddemål, LBK nr 115 af 31/01/2015 Gældende* (https://www.retsinformation.dk/Forms/R0710.aspx?id=167368).

14 *In line with Danish legislative practice, the section sign § is used whenever Danish legislation is referred to.*

important public policy portfolios as education, social affairs, health and research are striking. By attributing the lion's share of 81.24 per cent to the culture portfolio, the legal framework effectively takes the authorised gambling system into its ambit and makes its functions related to other public policy portfolios emerge as ancillary. Against this backdrop, it could still be argued (in an attempt to justify the system) that the ministry of culture has several important roles in its purview, with substantial related financial responsibilities. However the statutory instrument effectively earmarks the lion's share for the sports sector (63.83%) with the remainder being reserved for culture (18.56%), youth (9.79%) and a variety of other policy areas, each representing a few percent or even less (§ 3 LBK 115).

At this point one might be forgiven for expecting the sport portfolio budget to be distributed through open calls for proposals with evaluations of applications, transparency and equal treatment of applicants protected by administrative law. However, this is not the case, as the 63.83 per cent of the ministry's 81.24 per cent share is, in turn, distributed to designated beneficiaries who are *not even* required to apply. In 2015, these designated beneficiaries were defined by law as the elite sport promotion agency Team Danmark (9.91%), the two major (historically rivalling, culturally and politically distinct) sports federations DIF (Danmarks Idrætsforbund) (30.83%) and DGI (previously Danske Gymnastik- & Idrætsforeninger) (28.63%), the Danish Company Sport Federation (Dansk Firmaidrætsforbund) (4.25%), a national funding scheme for sports, cultural and recreational facilities (Lokale og Anlægsfonden) (9.20%), the national antidoping agency (Anti-Doping Danmark) (1.84%), the international sports events organization agency Sport Event Denmark (2.55%) and the Danish Institute for Sports Studies (Idrættens Analyseinstitut) (IDAN) (0.85%), respectively. Note the different identities of the centre-right sports confederation DIF, the centre-left sports confederation DGI and the 'company sports' federation DFIF. In contrast to many other countries, and as a consequence of its specific history, sport in Denmark at the national level is organized into three major umbrella organizations which have different aims, structures and ideologies. Only DIF is affiliated with the International Olympic Committee (IOC) and only DIF selects athletes for the Olympics, yet DIF is also a sport-for-all organization in much the same way as Germany's Deutscher Olympischer Sportbund (DOSB) and France's CNOSF (Comité national olympique et sportif français) are, being a joint umbrella organization for elite, amateur and recreational sports.[15] However, the history of DIF has stronger links with royalty, bourgeoisie, sponsors and competitions than that of DIF, whose roots stem from the years of provisional cabinet legislation under J. B. S. Estrup (1825–1913), a conservative landowner-politician who, between 1885 and 1894, aimed to turn back the clock of the then emerging democracy. As a result, gymnastics and shooting associations were established in rural areas, with the barely concealed aim of organizing cells which would have been able to organize resistance, should Estrup's nascent dictatorship have swapped over into a full-fledged one. DGI's identity is strongly linked with German and Swedish gymnastics,[16] and with non-competitive forms of sports and physical activity, though today's DGI is also a joint umbrella organization for diverse types of sports, albeit without the Olympic and the high-level amateur aspects found in DIF.[17] Finally, DFIF is an umbrella organizing various sports offered by employers in the workplace. The idiosyncrasies of these historical trajectories are deeply fascinating and not without recalling similar trajectories in nineteenth-century Britain which saw four types of sports clubs emerge: workplace-based, religious, political, and those developed by the alcohol trade

15 *"The National Olympic Committee and Sport Confederation of Denmark (DIF) is an umbrella organisation for 61 national sports federations, encompassing approximately 9000 local clubs and approximately 1,900,000 members. The organisation was founded in 1896 and became the central organisation for a wide range of sports federations and also the body in charge of organising Danish national championships. This led to natural ties with the Danish Olympic Committee, a body founded in 1905 and then merged with the DIF in 1993. The DIF governs both 'Sport for All' and 'elite sport' and at the national level the organisation is responsible for issues of common interest across the different sports federations" (Ibsen, 2017, p. 95).*

16 *"The Danish Gymnastic and Sports Associations (DGI) is the umbrella organisation for 15 regional associations (decentralised units of the DGI), which focus on the pursuit of 'Sport for All'. In 2015, the DGI has approximately 6000 clubs and 1,500,000 members. The origin of this organisation was the Danish Rifle Shooting Association, founded in 1861, which later included (Swedish) non-competitive gymnastics, and then evolved into the DGI. This part of Danish sport was originally a part of the agrarian movement in the rural areas of the country, with ideals of sport other than those of the Olympics (DGI, 2014)" (Ibsen, 2017, p. 95).*

17 *"The Danish Federation of Company Sport (DFIF) has 80 local company sports associations with approximately 370,000 members. The DFIF was founded in 1946 and played the lead in developing a particularly Danish tradition of merging the personal and professional spheres by engaging in sporting activities with colleagues [. . .]. The DFIF provides sport and physical activities for people during their affiliation to a workplace in approximately 8000 informal groups or clubs (DFIF, 2015)" (Ibsen, 2017, p. 96).*

(Vamplew, 2013), just like Borussia Dortmund was initially set up by Polish Catholic clergy in an attempt to create useful leisure activities for young male members of a recently migrated mining community.[18] What makes this Danish structure so unique (and surprising, when compared with the generally evidence-based policy making of that country) is that it has been able to persist and sustain itself right until now: hardly the kind of structure one would be able to successfully campaign to introduce, had it not already been extant. And yet, all three umbrella organizations are entitled to structural grants on a vast scale, yet it would be misleading to represent this arrangement as a pluralistic one *per se*: rather than allowing for a multitude of actors, it is one which preserves the established rights of three big actors, without requiring them to actually apply for the grants and without a need to justify their activities. Optimistically (and in line with the dominant ideological discourse), the absence of specific targets, even when grants have been paid, is not merely the expression of an absence of accountability: rather, it is perceived as the expression of a 'relationship between government and voluntary associations, which is based on a respect for autonomy, but also on trust that associations will act in the spirit of including all members of society' (Ibsen & Elmose-Østerlund, 2016, p. 56).

The remainder (11.94%) is reserved for 'other sports policy objectives', presumably on the basis of open calls (§ 3 a LBK 115). In effect, while the system does meet the requirements of legality and transparency, it could hardly be less evidence-based and, rather than allowing for an open and equitable distribution of public money to all interested civil society organizations, it provides a well-oiled machinery for serving vested interests in a way which may surprise foreign observers of a Nordic democracy. Intriguingly, while the statutory instrument provides detailed figures for fiscal 2015, subsequent provisions include updated figures for fiscal 2016. While one might have expected the 2015 figures to be based on membership statistics, which would at least allow for a minimum of evidence-based policy making (even if exclusive distribution to designated beneficiaries may seem objectionable from a public policy point of view), this cannot possibly apply to the 2016 exercise also which, in 2015, may only have been based on estimates or, by default, political bargaining. The statutory instrument further provides for the boards of certain beneficiaries to allocate their money to other beneficiaries (§ 3 a Stk. 13–17 LBK

18 *See, for example, http://www.porta-polonica.de/en-sysgb/Atlas-of-remembrance-places/polonia-dortmund*

115), and as these boards include members appointed by the minister of culture, a measure of openness and transparency is at least a distinct possibility in these cases. Redistribution of money by Team Danmark (§ 3 a Stk. 1 LBK 115), and Anti-Doping Danmark (§ 3 a Stk. 14 LBK 115), is subject to rules laid down in two specific acts of parliament. By contrast, the redistribution of money allocated to DIF, DGI and DFIF is explicitly at the discretion of these beneficiaries (§ 3 a Stk. 12 LBK 115). Grants allocated to 'other sports policy objectives' are subject to parliamentary scrutiny: this looks distinctly reassuring, but it should be remembered that this represents only 11.94 per cent. Conversely, 88.06 per cent of the money is distributed, by law, outside the scope of parliamentary scrutiny.

While many of these provisions may seem surprising, the casual observer might still be forgiven for assuming that administrative law and freedom of information would apply, *mutatis mutandis,* to all of these unusual arrangements, yet this is specifically ruled out by means of a provision inserted at the end of § 3. This provision explicitly restricts the applicability of the relevant acts of parliament[19] to grants allocated to Team Danmark, the national facilities scheme, Anti-Doping Danmark, Sport Event Denmark and IDAN (§ 3 a Stk. 19 LBK 115). This effectively means that only money paid to small organizations with less political clout is subject to administrative law and freedom of information guarantees, while it is also interesting to note that the (really marginal) amounts paid to meet 'other sports policy objectives', while subject to parliamentary scrutiny, are also considered not to be covered by such essential legal guarantees. The resulting exemptions all have the potential to seriously curb the potential of citizens to exercise their fundamental rights and for other (unaffiliated and unrecognised) civil society organizations to defend their legitimate interests. These provisions provide a telling example of legality without legitimacy. In view of future research, it would furthermore be interesting to compare the proportionate shares of funding attributed, by law, to DIF, DGI and DFIF and see if they match the respective membership figures. Similarly, it might be of interest to establish the extent to which the population's preferences for specific sport and HEPA activities is matched by the panoply of activities offered as part of the structures and activities funded through the funding streams described in this section of this chapter. How would the iconic status of handball in Danish society, as a spectator sport, match investment in handball infrastruc-

19 *Forvaltningsloven, Lov om offentlighed i forvaltningen.*

ture and activities offered, bearing in mind that much of this investment may actually be borne by local councils, possibly without any lottery-generated resources? And how would the outcome of this exercise, in turn, compare with possibly dissimilar popular preferences in cities and (semi-)rural areas, respectively? Would young people with an immigration background living in cities opt for handball, or rather for football (soccer)? Are public budgets, including lottery-based as well as taxation-based budgets, used in ways which broadly meet the diverse preferences of (all of) the inhabitants of the country?

More crucially, perhaps, there is no guarantee (and indeed no obligation) that the vast sums disbursed by central government to the three national sports confederations will benefit the local grassroots level at all, as such activities will most likely be funded by local councils (municipalities) out of budgets covered by general taxation.[20] Paradoxically, rather than making sport-for-all providers self-contained and independent, this system may actually have the effect of sheltering national sports organizations from the grass roots, as these two different layers within the sports movement are funded through funding streams which are entirely different and separated from each other.[21] Against this backdrop, one might even question the sport movement's claim that it represents one, unitary social movement.

---

20 *"There is almost no direct – or indirect – financial support from the state for local sports clubs – the main support comes from municipalities (local government), although enshrined in national legislation. Clubs benefit indirectly if they take part in activities of national sports organisations, which have been funded by central government. Local sports clubs can apply for money from a government fund to promote specific purposes, but this is insignificant overall" (Ibsen & Elmose-Østerlund, 2016, p. 56).*

21 *"But does it make any difference whether the state and the municipalities support sports organisations and clubs in one way or another? In relation to the government's ability to promote political objectives of the sport, the existing form of support on the national level – basic grants with very few requirements – gives small and indirect opportunities to pursue an active sports policy. The state support is neither connected with precise indications of what the money should be used for nor triggered by performance obtained. In turn, the current form of support is easy to manage and gives each sports organisation a great autonomy to pursue their own goals and interests. As a result of state aid, the sports organisations are independent from their own sports clubs – in economic terms – because the membership fees represent a very small proportion of the organisations' revenue. The current form of support is therefore neither ideal from the point of view of the political influence on how state support is used, nor from the perspective that the support must strengthen the purpose of organisations and their raison d'être: To work for their member clubs' interests and needs" (Ibsen, 2017, p. 107).*

14.2.3 Discussion

While the case study presented in the previous section is bound to trigger questions about the Danish sport policy model (at least as far as finance is concerned), the objective of this chapter is not to expose Denmark in particular. Rather, the fact that this model can be so easily exposed – due to the transparency involved when the relevant details are laid down in legislation – makes for a methodologically rather obvious case study. Critical examples of a mismatch between a perceived HEPA-friendliness and a continued pervasiveness of more traditional sport policy orientations and values could be extracted from other EU countries as well. Public investments in the Olympic Games of Athens and London, or the FIFA World Cup in Germany or the UEFA Euro in Portugal, have all fostered critical debates regarding the use of taxpayers' money (Kornbeck, 2015a). The ensuing realisation that the bidding system may be to blame, as it has driven public authorities to overbid each other far beyond the limits of fiscal responsibility, has been reflected in the work of UNESCO's World Conference of Sport Ministers (MINEPS V). On 30 May 2013, at MINEPS V held in Berlin, ministers in the so-called 'Declaration of Berlin' adopted by 121 represented member states, demonstrated an acute awareness of these developments,[22] while also acknowledging the multifarious and nefarious impacts on national and local economies, including on the capacity to pursue responsible public policies.[23] On this basis, MINEPS

---

22 *MINEPS V. Declaration of Berlin,* http://unesdoc.unesco.org/images/0022/002211/221114e.pdf, *p. 6: "2.10 Being aware that major sport events are subject to continually increasing financial, technical and political requirements that may act as a disincentive to bid for major sport events and risk excluding certain countries from the bidding for or hosting of such events".*

23 *Ibid., p. 7: "2.15 Acknowledging the data which shows that many oversized stadia are not financially viable post-events (while generating maintenance costs);*

2.16 Noting the trend of increasingly competitive bids and 'overbidding', that is, incurring higher costs than necessary in order to outbid competitors, by countries wishing to host major sport events, and a corresponding escalation of hosting costs, which are frequently underestimated in ex-ante studies;

2.17 Recognizing that political and financial support by the public sector, and its early involvement, are prerequisites for the organisation of major sport events as of the bidding stage;

2.18 Recognizing that participation in the bidding process for hosting a major sport event and the related international exposure can act as a catalyst for sustainable national development, improved cooperation of different societal groups and identity building;

2.19 Emphasizing the importance of transparent community participation in the bidding and implementation process for major sport events to avoid undesired changes in the living environment of local residents, including the displacement of local populations and subsequent gentrification."

delegates committed themselves to counteract these problematic trends, thereby demanding new bidding procedures.[24] If based on previous experience[25] critical observers might be sceptical about the potential for this MINEPS initiative to trigger real change, including because NGOs rather than governments are the owners of the said bidding procedures, at least it could be said that the intentions seemed excellent. The information available on the follow-up to MINEPS V,[26] including the subsequent MINEPS VI conference (Kazan, 2017)[27] and the Global Action Plan (2018)[28] linked to the same, do however not provide sufficient information to judge about real-term policy change in the countries involved.

If a sea change is to be expected as regards the bidding procedures operated by international sports organizations (IOC, FIFA, etc.) in view of major international competitions, it seems far more likely that this change will come from sports organizations themselves, as opposed to being initiated by governments. The fact that the expectations of the IOC, for instance, have been able to increase continuously since the 1980s, should be considered against the backdrop of a widespread willingness to bid – until the post-2008 crisis started to 'bite', that is. The German embarrassment cause by the 2015 Hamburg referendum

---

24 *Ibid., p. 8: "2.27 Treat major sport events as an integral part of national physical education and sport planning, ensuring that other programmes do not suffer from budget shifts in favour of the implementation of major sport events or of high-performance sport;*

2.28 Commit, when hosting major sports events, to the sustainability of sport infrastructure for physical education, sport for all and high-performance sport and other community activities, in order to ensure that all concerned stakeholders can participate in and benefit from such events;

2.29 Develop a consistent policy setting out the conditions for planning and implementing major and mega sport events as well as for participating in related bidding procedures."

25 *Prior to MINEPS V, one expert author noted: "In line with the UN's Convention on the Elimination of All Forms of Discrimination against Women (adopted in 1979) UNESCO has sought to promote equity in the development of national and local sporting projects and programmes. Although this is an aspect of providing moral leadership, few practical projects are associated with this goal beyond the establishment of an Observatory on Women and Sport following the 2004 Conference of Ministers (MINEPS IV) which took place in Athens. Impacts are thus difficult to gauge" (Henry, 2010, p. 276).*

26 *http://www.unesco.org/new/en/social-and-human-sciences/themes/physical-education-and-sport/mineps/follow-up-to-mineps-v/*

27 *http://unesdoc.unesco.org/images/0025/002527/252725E.pdf*

28 *https://en.unesco.org/news/ministerial-conference-sport-and-physical-education-adopts-global-action-plan-promote-safe*

was not the only one in its kind. Indeed, only the year before the Rome bid for the 2024 Games had been withdrawn because the public authorities had been reluctant to underwrite the bid on account of the crisis. When on 13 September 2017, the IOC announced its simultaneous decision to award the 2024 Games to Paris and the 2028 edition to Los Angeles, its deliberately upbeat communication style could not conceal the awkwardness of the situation. For initially Paris, Hamburg, Budapest, Rome and Los Angeles had been the five candidate cities bidding for the 2014 Games. But as three bidders had withdrawn, the IOC was left with only two and decided to award each of the candidates its own edition of the Games. The overflowing enthusiasm expressed in the speech of IOC's Thomas Bach[29] could not conceal the objective reality that the IOC suddenly had had far less choice than in the past. Press pictures showing Paris mayor Anne Hidalgo with tears in her eyes also looked slightly overplayed, given that the only alternative outcome would have been for Paris to take its turn after Los Angeles rather than before.

Even the most enlightened countries, which are otherwise often (including by themselves) seen as role models to other countries, may sometimes pursue surprisingly traditional, HEPA-irrelevant sports policies, competing for the right to host and for opportunities to harvest medals. A powerful illustration hereof can be seen in recent statements made by German sports minister Thomas de Maizière, announcing an expectation of 30 per cent more medals in return for public investment in elite sports. Apart from the fact that such an objective may seem objectionable at a time when Germany is simultaneously committed to reducing doping in sport, including by passing one of the most draconian anti-doping laws found in any Western liberal democracy,[30] the rhetoric used by the minister places such high value on elite sports that critical observers must wonder what are the implications

29 *IOC makes historic decision by simultaneously awarding Olympic Games 2024 to Paris and 2028 to Los Angeles. Date 13 Sep 2017. https://www.olympic.org/news/ioc-makes-historic-decision-by-simultaneously-awarding-olympic-games-2024-to-paris-and-2028-to-los-angeles: "Congratulations to Paris 2024 and Los Angeles 2028! This historic double allocation is a 'win-win-win' situation for the city of Paris, the city of Los Angeles and the IOC," said the IOC President, Thomas Bach. 'It is hard to imagine something better. Ensuring the stability of the Olympic Games for the athletes of the world for the next 11 years is something extraordinary," Bach commented.'*

30 *Anti-Doping-Gesetz (AntiDopG). (Anti-Doping-Gesetz vom 10. Dezember 2015 (BGBl. I S. 2210), das zuletzt durch Artikel 6 Absatz 5 des Gesetzes vom 13. April 2017 (BGBl. I S. 872) geändert worden ist. This act of parliament foresees prison sentences up to three for doping (not just for trade in doping substances, facilitation, instigation, etc.).*

for HEPA policies. In an interview published in 2015, Mr de Maizière referred to medals as 'the currency', emphasising that for Germany to gain more of these, he would recommend drawing inspiration both from the tradition of former West Germany and also from that of former East Germany. Based on these combined traditions, 'and given the size of our economy and the volume of our investments in top sports, we should be able to achieve a third more medals, perhaps even more'. Mr de Maizière's positive reference to the former East German tradition was rendered the more intriguing by the fact that the question prompting this statement was one in which the journalist referred explicitly to that country's use of state-sponsored doping to harvest more medals. Asked if such an objective would not necessitate additional financial investment, the minister replied that, on the contrary, it was already so high that 'German tax payers are entitled to a return on investment'.[31]

Against the backdrop of such inertia in public policies, the physical activity guidelines presented earlier (section 14.1) offer an opportunity to critically revisit both the Danish case study (section 14.2) and also the example of Mr de Maizière (section 14.3). While at the EU level, an emerging EU HEPA doctrine can be identified, including in the case law of the EU courts (Kornbeck, 2018), a similar change of mindset seems to take some time at the national level. The Danish case study certainly does not meet the standards defined in the physical activity guidelines. But if these guidelines are being used successfully as benchmarks for funding (like in Erasmus+) (European Commission, 2017), they definitely can have an impact, there being no conceptual impediment to their use in national funding policies. The EU has been able to use the guidelines as a gold standard, driving a conceptual change in a field where the EU has no power to harmonise national rules (Article 165 TFEU), its mandate being limited to an EU competence to 'support, coordinate or

31 *„Was genau ist das Ziel? Da muss man nicht darum herum reden. Medaillen sind die Währung. Das heißt nicht, dass nicht auch ein knapper vierter Platz etwas Wertvolles ist und dass ein fünfter Platz im Endlauf über 100 Meter vielleicht mehr wert ist als manche Gold- oder Silbermedaille in einer deutschen Domäne. Aber wenn wir den Spitzensport fördern, dann geht es eben nicht um gehobenen Breitensport. Und Erfolg oder Misserfolg im Spitzensport misst sich an Ergebnissen. Ist denn Ihr Maßstab wirklich ein Ergebnis wie das der DDR bei den Anabolika-Spielen in Seoul 1988? Zweiter vor den Vereinigten Staaten?*

Nein, absolut nicht. Und wir müssen auch nicht den Rechenschieber anlegen. Aber wir müssten eigentlich nach der Tradition in beiden deutschen Staaten und nach unserer Wirtschaftskraft, mit der wir den Spitzensport fördern, mindestens ein Drittel mehr Medaillen bekommen, vielleicht mehr. Kostet das nicht noch mehr Geld? Der Steuerzahler hat für das Geld einen Anspruch auf Gegenleistung. Es liegt in unserem Interesse und auch im Interesse des DOSB, für das gleiche Geld mehr herauszukriegen." (Simeoni, 2015)

supplement the actions' of member states (Article 6 TFEU). But while the precise impact of certain HEPA-friendly references enshrined in the sport article of the treaty (Article 165 TFEU) cannot be seen as having decisively turned the tables in judgments such as *Murphy*[32] and *Bernard*,[33] both of which ruled on matters strictly limited to professional sports, the same article has in fact played a useful role in informing the commission's and the court's decisions on the compatibility of state aid with the single market (Article 107 TFEU). Here, the degree to which facilities and services supported through state aid actually benefit the population at large, as established in *Hamr Sport*[34] and *Magic Mountain Kletterhallen*,[35] does seem to provide a yardstick for public policy and legal interpretation with HEPA-friendliness as the cornerstone. If the idea is to focus on exempting specific activities, as opposed to specific legal entities (Kornbeck, 2017b), then these recent developments suggest a convergence between the application of EU competition law (hard law) and the promotion of a HEPA policy agenda (soft law), which justifies speaking of an 'emerging doctrine' (Kornbeck, 2018). This in turn would suggest that alternatives to the traditional sports policy models pursued by some member states do exist.

14.3 Conclusion

Methodologically speaking, the Danish case study discussed earlier in this chapter is not without its limitations, and indeed it would be welcome if future research could shed light on the situation in other EU member states. It does however seem fairly certain that revenue generated by the (only partially liberalised) state gambling monopolies does not

---

32 *Joined cases C-403/08 and C-429/08 Football Association Premier League Ltd & Karen Murphy [2011] ECR I-09083.*

33 *Judgment of the court (Grand Chamber) of 16 March 2010. Olympique Lyonnais SASP v Olivier Bernard and Newcastle UFC. Case C-325/08. ECR 2010 I-02177. ECLI:EU:C:2010:143.*

34 *Judgment of the General Court (Eighth Chamber) of 12 May 2016. Hamr Sport a.s. v European Commission. Case T-693/14. Electronic reports. ECLI:EU:T:2016:292. Judgment of the General Court (Sixth Chamber) of 9 June 2016.*

35 *Judgment of the General Court (Sixth Chamber) of 9 June 2016. Magic Mountain Kletterhallen GmbH and Others v European Commission. Case T-162/13. Electronic reports. ECLI:EU:T:2016:341.*

always reach the grass roots: indeed, one study found that HEPA activities are, more often than not, paid by local councils or by private households themselves (Eurostratégies, Amnyos, CDES & DSHS, 2011, pp. 46, 80). At any rate, the financial circuits need to be far better understood than is presently the case (Kornbeck, 2015a). Specifically, even if the Danish lotto system may seem generous in total numbers, the €122 million distributed by the ministry of culture, in the fiscal year 2013, to sports organizations (mainly to DIF and DGI) was dwarfed by the €425 million distributed, in that same year, by local councils, largely to pay for facilities open to all citizens, rather than to support clubs (Ibsen, Elmose-Østerlund & Laub, 2015, p. 94), raising the question whether ordinary citizens really profit from the presence of such organizations, if their focus is rather on sport for all. With debates on the impact of austerity policies being widespread not only in the media, but indeed also in sports-related scholarship, intriguing papers have been contributed, not only (and predictably) from the United Kingdom (Brown & Pappous, 2017; Kenyon, Mason & Rookwood, 2018; Ramchandani, Shibli & Kung, 2018; Reid, 2018; Rossi & Jeanes, 2017; Widdop, King, Parnell, Cutts & Millward, 2017) and (even more predictably) from Greece (Papadimitriou & Alexandris, 2017), but also from such countries as Belgium (Haudenhuyse, 2018), Denmark (Bundgaard Iversen, 2017) and the Netherlands (Hoekman, Roest & Poel, 2017). The Greek paper even indicated that there was no more funding *elite* athletes (Papadimitriou & Alexandris, 2017). Some papers dealing with municipal counter-policies, including Liverpool (Kenyon, Mason & Rookwood, 2018) and Edinburgh (Reid, 2018), not only smack of the 'municipal socialism' practised in the first half of the twentieth century: implicitly they vindicate the HEPA paradigm and validate the EU Physical Activity Guidelines. In so doing, they (equally implicitly) focus on 'le geste sportif', while they may be interrupting the grand narrative of traditional sports policy, epitomised by the motto preceding this chapter.

References

Brettschneider, W. D., Hoffmann, D., Naul, R., & Steinzen, A. (2009). *Von den Erkenntnissen aus dem EU-Bericht „Young people`s lifestyle and sedentariness" zur handlungsorientierten Umsetzung der Empfehlungen im Rahmen von zwei lokalen Präventionsstudien.* In R. Naul, A. Krüger, & W. Schmidt (Eds), *Kulturen des Jugendsports. Bildung, Erziehung und Gesundheit* (pp. 67–104). Aachen: Meyer & Meyer.

Brettschneider, W. D., & Naul, R. (2005). *Study on young people's lifestyles and sedentariness and the role of sport in the context of education and as a means of restoring the balance*. Retrieved from http://www.uni-muenster.de/imperia/md/content/sportwissenschaft/eu-studies/abstracts/noored_ja_sport_04_lotpaderborn.pdf

Brown, C., & Pappous, A. (2017). The organisational performance of national disability sport organisations during a time of austerity: A resource dependence theory perspective. *International Journal of Sport Policy and Politics*. Published online: October 19, 2017. https://doi.org/10.1080/19406940.2017.1381635

Bundgaard Iversen, E. (2017). Public management of sports facilities in times of austerity. *International Journal of Sport Policy and Politics*. Published online: January 20, 2018. https://doi.org/10.1080/19406940.2018.1426621

Eureval-C3E. (2005). *External evaluation (ex-post) of the European Year of Education through Sport EYES 2004*. Retrieved from http://ec.europa.eu/dgs/education_culture/more_info/evaluations/docs/sport/eyes_en.pdf

European Commission. (2008). *Brussels, 10 October 2008. EU physical activity guidelines. Recommended policy actions in support of health-enhancing physical activity*. Approved by the EU Working Group "Sport & Health" at its meeting on September 25, 2008. Retrieved from http://ec.europa.eu/assets/eac/sport/library/policy_documents/eu-physical-activity-guidelines-2008_en.pdf

European Commission. (2017). *Erasmus+ Programme Guide. Version 2 (2017)*: Retrieved from: http://ec.europa.eu/programmes/erasmus-plus/sites/erasmusplus/files/files/resources/erasmus-plus-programme-guide_en.pdf

Eurostratégies, Amnyos, CDES [Centre de Droit et d'Economie du Sport] & Deutsche Sporthochschule Köln. (2011). *Study on the funding of grassroots sports in the EU. With a focus on the internal market aspects concerning legislative frameworks and systems of financing*. Final report. Vol. I. Brussels: Eurostratégies. Retrieved from http://ec.europa.eu/internal_market/top_layer/docs/FinalReportVol1_en.pdf

García, B., Vermeersch, A., & Weatherill, S. (2017). A new horizon in European sports law: The application of the EU state aid rules meets the specific nature of sport. *European Competition Journal, 13*(1), 28–61.

Geeraert, A., & Drieskens, E. (2017). Normative market Europe: The EU as a force for good in international sports governance? *Journal of European Integration, 39*(1), 79–94.

Haudenhuyse, R. (2018). The impact of austerity on poverty and sport participation: Mind the knowledge gap. *International Journal of Sport Policy and Politics*. Published online: January 19, 2018. https://doi.org/10.1080/19406940.2017.1406975

Henry, I. (2010). Sports development and adult mass participation: The role of international organisations. In B. Houlihan & M. Green (Eds.), *Routledge handbook of sports development* (pp. 267–280). Abingdon et al.: Routledge.

Hoekman, R., Roest, J. W. van der, & Poel, H. van der (2017). From welfare state to participation society? Austerity measures and local sport policy in the Netherlands. *International Journal of Sport Policy and Politics*. Published online: October 24, 2017. https://doi.org/10.1080/19406940.2017.1381636

Ibsen, B. (2014). Grundstøtte eller præstationsstøtte: Virkningen af forskellige former for statsstøtte til idrætsorganisationerne. In K. Eskelund & T. Skovgaard (Eds.), *Samfundets idræt: Forskningsbaserede indspark i debatten om idrættens støttestrukturer* (pp. 13–54). Odense: Syddansk Universitetsforlag (University of Southern Denmark Studies in Sport and Movement; 3).

Ibsen, B. (2017). Denmark: The dissenting sport system in Europe. In J. Scheerder, A. Willem, & E. Claes (Eds.), *Sport policy systems and sport federations: A cross-national perspective* (pp. 89–112). Basingstoke: Palgrave.

Ibsen, B., & Elmose-Østerlund, K. (2016). Denmark. In B. Ibsen, G. Nichols, & K. Elmose-Østerlund (Eds.), *Sports club policies in Europe. A comparison of the public policy context and historical origins of sports clubs across ten European countries* (pp. 55–58). Odense: Centre for Sports, Health and Civil Society, University of Southern Denmark.

Ibsen, B., Elmose-Østerlund, K., & Laub, T. (2015). Sport clubs in Denmark. In C. Breuer, R. Hoekman, S. Nagel, H. van der Werff (Eds.), *Sport clubs in Europe: A cross-national comparative perspective* (pp. 85–110). Cham (Switzerland) et al.: Springer.

Kenyon, J. A., Mason, C., & Rookwood, J. (2018). Emerging third-sector sports organisations and navigating uncertainty in an 'era of austerity': A single ethnographic case study from Liverpool. *International Journal of Sport Policy and Politics*. Published online: January 23, 2018. https://doi.org/10.1080/19406940.2018.1425732

Kornbeck, J. (2006). Governance als Soft Law: Innovation oder Notwendigkeit im sportpolitischen Handeln der EU? In W. Tokarski, K. Petry, & B. Jesse (Eds.), Sportpolitik. In *Theorie- und Praxisfelder von Governance im Sport*. Cologne: Sportverlag Strauß (Veröffentlichungen der Deutschen Sporthochschule Köln), *15*, 31–52.

Kornbeck, J. (2015a). Lisbonisation without regulation: Engaging with sport policy to maximise its health impact? *International Sports Law Journal, 15*(1–2), 112–122.

Kornbeck, J. (2015b). The diversity of sports volunteering in the European Union: Sharing information and good practice to provide EU added value in a field without EU regulatory powers. In N. Rosendal Jensen (Ed.), *Voluntary work and youth unemployment* (pp. 99–118). Copenhagen: Aarhus Universitet (DPU).

Kornbeck, J. (2017a). Bosman and athlete welfare: The sports law approach, the social policy approach, and the EU Guidelines on Dual Careers. *Liverpool Law Review, 38*(3), 307–323.

Kornbeck, J. (2017b). « Le geste sportif »: ein alternatives Unterscheidungsmerkmal zu den "besonderen Merkmalen" des Sports nach Artikel 165 AEUV? *Zeitschrift für europäisches Wirtschaftsrecht (EuZW) (European Business Law Review), 28*(15), 603–606.

Kornbeck, J. (2018). Health-friendly sport policy: An emerging soft law doctrine? In R. Parrish, B. Garcia, & J. Anderson (Eds.), *Research handbook on EU Sports Law* (pp. 49–78). Cheltenham, Gloucester, UK: Edward Elgar Publishing.

Papadimitriou, D., & Alexandris, K. (2017). 'Adopt an athlete for Rio 2016': The impact of austerity on the Greek elite sport system. *International Journal of Sport Policy and Politics.* Published online: December 26, 2017. https://doi.org/10.1080/19406940.2017.1416487

Ramchandani, G., Shibli, S., & Kung, S. P. (2018). The performance of local authority sports facilities in England during a period of recession and austerity. *International Journal of Sport Policy and Politics.* Published online: January 8, 2018. https://doi.org/10.1080/19406940.2017.1420676

Reid, G. (2018). Managing budget cuts in Edinburgh's sport and recreation services: Progressive localism in a resilient local authority? *International Journal of Sport Policy and Politics.* Published online: January 23, 2018. https://doi.org/10.1080/19406940.2018.1427127

Richter, C., & Naul, R. (2009). Versagensangst und Selbstvertrauen als psycho-mentale Merkmale in Relation zur Fußball-Fitness bei C- und D-Junioren des Fußball- und Leichtathletikverbandes Westfalen (FLVW). In U. Frick (Ed.), *Fußball in Schule und Verein* (pp. 99–106). Hamburg: Czwalina.

Rossi, T., & Jeanes, R. (2017). Is sport for development already an anachronism in the age of austerity or can it be a space of hope? *International Journal of Sport Policy and Politics.* Published online: October 25, 2017. https://doi.org/10.1080/19406940.2017.1380682

Siedentop, D. (1994). *Sport education: Quality PE through positive sport experiences*. Champaign, IL: Human Kinetics.

Simeoni, E. (2015). *Innenminister de Maizière: "Mindestens ein Drittel mehr MedaillenÐ*. Aktualisiert am 17.07.2015-13:08. Retrieved from http://www.faz.net/aktuell/sport/sport-politik/thomas-de-maiziere-im-interview-ueber-den-deutschen-sport-13707649.html?printPagedArticle=true#pageIndex_0

Vamplew, W. (2013). Theories and typologies: A historical exploration of the sports club in Britain. *International Journal of the History of Sport, 30*(14), 1569–1585.

Widdop, P., King, N., Parnell, D., Cutts, D. & Millward, P. (2017). Austerity, policy and sport participation in England. *International Journal of Sport Policy and Politics*. Published online: August 14, 2017. https://doi.org/10.1080/19406940.2017.1348964

Info

Recurrent abbreviations

CFI: Court of First Instance (now the General Court)

Court: Court of Justice of the European Union (CJEU) (previously Court of Justice of the European Communities)

EYES: European Year of Education through Sport 2004

HEPA: health-enhancing physical activity

TEC: Treaty establishing the European Community (TEC) (Nice Treaty) (defunct)

TEU: Treaty on European Union (Lisbon Treaty)

TFEU: Treaty on the Functioning of the European Union (Lisbon Treaty)

PART FIVE

SPORT POLICY IN THE LOCAL CONTEXT

15 Weight status of children with a Turkish or Syrian background in Germany and the Netherlands; interventions such as Simple-Fit! and GKGK seem necessary

Monique L'Hoir, Sunday Nuraydin and Dennis Dreiskämper

This chapter discusses whether parents and children with a Turkish or Syrian background should be offered more preventive programs that include nutritional status, physical activity, and weight development. For Syrian children two parallel developments provide evidence to do so: (1) The Barker hypothesis, partly derived from pregnant women during the Dutch hunger winter. UNICEF and Amnesty International clearly indicate that Syrian refugees, living in camps or fleeing from occupied cities, are often malnourished, which legitimizes the comparison with the Barker hypothesis. The second parallel development is a group of children in the Netherlands and Germany who are at high risk for overweight and obesity, which is known for quite some time and certainly needs more attention (Schönbeck et al., 2011; De Onis & Blossner, 2000; Ng et al., 2014; Van Dommelen et al., 2015). We refer to children with a Turkish background. In addition to monitoring and identification of overweight in young children by the preventive youth health organizations and/or paediatricians, early intervention is necessary. Chain partners should join forces. A 'wait and see' policy is not an option for this group of vulnerable children. An intervention for young children was proposed and data of the longitudinal intervention study 'Healthy Children in a Sound Community' (GKGK) that includes children in of the age of six to ten is described. This study, that included 689 German and Dutch children and 83 children with a Turkish background, shows that by daily physical activity and weekly lessons on healthy food at school, after three years of intervention, mainly girls with a Turkish background, reduced their BMI, increased their weekly time of physical activity, and improved their motor abilities – especially aerobic endurance capacity – although they had high screen/media time levels. Furthermore, they were well integrated in their classes and had a stable, positive self-concept.

15.1 Introduction

Children with a Turkish or Syrian background are at higher risk for obesity compared to German and Dutch children.* Data from infants and toddlers have been collected as well as data from children at elementary schools (GKGK). Regarding children with a Syrian background and their adaptation in Germany and the Netherlands in relation to their weight development, less information exists. Therefore, in this chapter we draw a parallel between these two populations. Syrian children and their parents come from a war zone, and through different procedures they are able to settle in communities in Germany and the Netherlands. They receive a residence permit on humanitarian grounds. Usually they are medically examined and get a food intake by a health nurse or a physician. Vitamin deficiencies and illnesses are diagnosed. Shortly after their arrival, overweight, or obesity are not often present. The question is whether overweight lies ahead in these children? Do we need to start with preventive advice for these families, shortly after arrival? In this chapter we plead to do so, based on important facts and findings.

15.2 Refugees from Syria

In the Netherlands in 2015, about 30,000 Syrian refugees were housed in different municipalities (Pharos fact sheet: Syrian refugees, 2015). At the end of 2015, more than 360,000 Syrian citizens (combined 118,196 as of December 2014 and 243,721 as of October 2015) resided in Germany (German Interior Ministry, 2015). In their homeland, pregnant women did not receive prenatal care, and chronic conditions were difficult to treat or not treated at all because of the lack of, and high prices of, medications. Families who arrived in Western countries may have limited health skills, and their German or Dutch language often is insufficient. The official language in Syria is Arabic, but often people speak also Kurdish and Armenian. Higher-educated Syrian refugees often speak English and French. Of the Syrian refugees who come to the Netherlands, it has been estimated that about 5 percent has an academic education. There is uncertainty about the average level of education of the Syrian refugees (Centre for Asylum Seekers, COA). Since 1981, compulsory schooling up to the age of 15 has been established in Syria. Children aged 6 to 12 years attend primary school. From 12 to 15 years they go to a 'middle school'. Before the war in March 2011, 97 percent of the Syrian children went to primary school and 67 percent to high school. The religious background of Syrian children, and religion-related food standards, may differ (table 15.1).

Table 15.1: Religious groups in Syria

	(%)
Christian	10
Islamitic-Soennitic	74
Islamitic, other (Shiites, Alawites, Ishmaelites and Yazidi)	13
Druze	3

15.3 Two developments that may affect childhood obesity

Two risk-increasing factors for obesity are: (1) Previous long-lasting malnutrition in pregnant refugees and (2) unhealthy eating and drinking by Syrian parents and children after arrival in Western Europe, regardless of their nutritional condition on arrival in in Germany or the Netherlands.

The first development is knowledge that has been gathered following the children who were born in the hunger winter in the Netherlands in 1944, the so-called Barker hypothesis or hypothesis of the foetal origin of diseases in adulthood. This hypothesis states that, amongst other things, Type 2 diabetes mellitus and hypertension, insulin resistance and insulin deficiency, at least in part, result from adjustments of the foetus to malnutrition. Studies based on the Barker hypothesis show that a rapid weight growth in the first months of life, regardless of the birth weight of the child, in itself increases the risk of overweight in adulthood (Hales & Barker 1992; Oudhof et al., 2013). In non-industrialized countries the combination of a low birth weight, an unhealthy diet, and obesity in adulthood seems to result in a much larger increased risk of heart disease, hypertension, and diabetes mellitus Type 2 than in Western countries (White, 2000). Fetal malnutrition and rapid weight gain in the first months of life therefore form a great risk for developing obesity later in life.

The second development is the increased risk of overweight in children with a Turkish background who live in Germany or the Netherlands. National figures show that the prevalence of overweight in children already at a very young age with a Turkish background is higher than that in originally Dutch children (table 15.2).

Table 15.2: Prevalence of overweight and obesity in children with a Turkish background and Dutch children aged two and three years in percentages (Dommelen et al., 2015)

Age	2 Years		3 years	
Background	Turkish	Dutch	Turkish	Dutch
	(%)	(%)	(%)	(%)
Boys, overweight	25	8	26	7,8
Boys, obese	7	1,6	8	0,8
Girls, overweight	17	8,3	20	12,8
Girls, obese	4	0,7	5	0,7

Syria is adjacent to Turkey, and both countries have cultural similarities because of their shared history (Ottoman Empire), which is expressed in religious celebrations, daily lifestyle, food culture, education, and hospitality. Children from these areas who now live in the Netherlands or Germany are comparable in many aspects. Observations indicate that Syrian families allow their children to consume sweets, cakes, and soft drinks unlimitedly. This attitude is anchored in the culture. The young child gets what it wants, when it asks for. It is taboo for parents to refuse something to their child. Syrian culture is just like Turkish culture: a culture of excesses concerning food. Parents with a Turkish background eat three hot meals a day. And the kids of course do the same. In this sense, Turkish culture is similar to Syrian culture (Nuraydin, 2016).

One in 240 children in the Netherlands is morbidly obese; amongst children with a Turkish background, this is 1 out of 40 (Dommelen et al., 2014). This prevalence of overweight and obesity in young children with a Turkish background, and the large differences with the Dutch prevalence of morbidly obesity in children, is alarmingly high, and it is to be expected that within a few years, overweight and obesity will increase, also in Syrian children. Cultural differences seem to play a central role. Child rearing styles of Turkish parents differ from Dutch parents. Turkish parents have difficulty in setting limits and reward their children often with candy. There are indications that not only authoritarian, but also permissive, child rearing styles are associated with an increased risk of overweight (Factsheet GGD, http://www.kvnog.nl/factsheets). Dutch research shows that infants of parents with a Turkish background daily get an unnecessary extra bottle of milk, which is not offered to Dutch children (Eijsden et al., 2015).

15.4 Determinants of overweight

There is little reason to believe that the research findings among parents with a Turkish background or other non-Dutch ethnic background from the same region does not hold true for parents and children with a Syrian background. The risk of overweight is influenced by several determinants including cultural, religious, and socio-economic factors. A low socio-economic status (SES) explains only part of the difference in prevalence rates of overweight and obesity between Dutch children and children with a Turkish background. The prevalence rates of obesity are higher in children with a Turkish background with a low SES than in Dutch children with low SES (Dommelen et al., 2015).

Education and parenting skills are important determinants of overweight in young children, which was a result of Dutch research carried out in 390 parents of children from infancy to four years old (Boere-Boonekamp et al., 2008). Parents with a non-Western background, parents who are unemployed and parents with a low socio-economic status are more at risk of increasing child-rearing behaviours that are related to overweight than other parents (Boere-Boonekamp et al., 2008). Examples include giving soft drinks if the child asks for it and allowing the child to choose when and how long it watches TV. Learning early to eat breakfast and reducing or abandoning drinking fruit juice or sweetened drinks are promising interventions in young children (less than four years), precisely because this behaviour is least prevalent in low-educated and non-Western parents (L'Hoir et al., 2006).

Structural determinants, such as education, income class, and job also play a significant role. A relatively large proportion of parents with a Turkish background belongs to the low- or middle-educated categories. Beenackers et al. (2015) describe that 'intermediate determinants also influence health, such as financial situation, psychosocial factors, behavioral factors, biological factors, and social cohesion.' For parents with a Turkish background, and likely also for Syrian refugees, all these unfavourable factors count, except the last. There is a high degree of social cohesion in Turkish and Syrian communities.

Excessive overweight of the mother is an important determinant of obesity (Whitaker et al., 1997). On an average, 17 percent of the difference in obesity in two-year old Dutch children and children of the same age with a non-Western background is determined by the amount of the self-reported BMI of the mothers (24.6 (SD ± 4.6) before they become pregnant, and 33 percent is determined by the average increase in weight in the first six months of the baby after the birth (De Hoog, 2012). Recent studies show that due to the unhealthy lifestyle of mothers during pregnancy, the DNA methylation in the cord blood of their children can change, and as a result these children are programmed as vulnerable to an unhealthy 'obesogene' environment later in life. The period from conception to two years is of great importance for the prevention of overweight (Küpers, 2016).

Unhealthy food and exercise patterns are more common in children with a Turkish background than in children with a Dutch background. The Dutch national growth study showed that in children aged two to four years old with a non-Western background (Turkish background *n* = 883) there was a high prevalence of the intake of more than two glasses of carbonated or fruit drinks (50%), less than one hour of physical activity (39%), and no walking or biking to childcare/preschool (28%) (Van Dommelen et al., 2015). This can be partly explained by the fact that mothers with a Turkish background more often perceive a 'full baby' as beautiful than Dutch mothers (43% vs. 22%) and are more often concerned about the growth of the baby (13% vs. 4%) (Van Eijsden et al., 2015). Mothers with a Turkish background more often breastfeed their child until at least six months OR ([OR] 3.0; 95% CI 1.6–5.6) but introduce complementary feeding more often before the age of six months. Children with a Turkish background on average receive one bottle of artificial milk per 24 hours more than Dutch children. Furthermore, when parents start with solid feeding before the age of six months, this more often is unusual food, such as yogurt and biscuits (OR 2.1; 95% CI 1.1–3.9) (Van Eijsden et al., 2015).

Also, young children with a Turkish background are offered fruit juice and syrup to drink more often, and they receive different kinds of snacks more often. The difference in perception of the weight of the baby (whether parents perceive their baby as overweight or having a normal weight) determines the difference in the food practices, but only for small food portion (Van Eijsden et al., 2015). When the baby cries excessively, the child is usually fed in response to the crying (Hulsmann et al., 2005). Recently it has been shown that overfeeding with breastfeeding is also possible (Willik et al., 2015). A short sleep duration is a determinant whose importance is becoming increasingly clear (Iglowstein et al., 2003; Monasta et al., 2010). Since in the Turkish culture children often go to sleep when they are tired (and first fall asleep on the couch, for example, which can happen late in the evening), this may disrupt the leptin and ghrelin levels, which are hormones associated with suppression of food intake and saturation. Excessive media use is associated with shorter sleep duration, because the blue light of an iPad postpones the production of melatonin, causing the child to get tired later and end up sleeping shorter (Landsmeer and others, 2014; Nji, 2015). Having a television in the bedroom also increases the risk of obesity.

15.5 Interventions

All pregnant women who come from Syria, should be interviewed about their environment and possible malnutrition. All young Syrian parents should receive preventive advice regarding the risk of overfeeding and get information about the relation between former malnutrition and current affluence. Malnutrition as well as overweight of the mother before she got pregnant and rapid growth of the baby in the first three months after birth should always give rise to active intervention. Health nurses and physicians should not wait with intervening till the potential overweight child reaches the age of three. The current policy for the prevention of overweight in children of parents with a non-Western background, and thus also Syrian parents, should be started early, in fact, shortly after arrival in the Netherlands or Germany.

Early intervention by enhancing the child-rearing skills of parents and including schools, municipalities and sports clubs seems more effective than intervening when children have reached their adolescence (L'Hoir and others, 2008, Yavuz and others, 2015, Bocca and others, 2011; Corner of, 2015). In children with a non-Western background it is often signalled too late that the child is gaining weight fast. This is to be expected to be similar in Syrian refugees and weight problems will also arise. Parents of infants and toddlers should be supported in developing their parenting skills so that they are able to apply the provided advice they get from the health nurse, physician or paediatrician for their children. Recently de Jong-Lenters (2016) demonstrated how important family factors and parenting skills are not only in relation to overweight, but also in relation to the development of caries. Shortly after Syrian refugees with young children arrive in the Netherlands or Germany, they should receive information about a healthy diet and get support in developing positive parenting skills. Since risk-increasing lifestyle behaviour starts shortly after birth of a child, intervention before and during pregnancy as well as immediately after birth is desired.

Cooperation between various care providers is a prerequisite for adequate prevention in overweight in children with Turkish or Syrian backgrounds. These are general practitioners, whom parents with a non-Western background and/or low SES prefer to consult for their problems (Bukman et al., 2016), healthcare nurses and physicians, who have the default contact moments with the families, dietitians who are able to inform Syrian parents about nutrients, vitamins and healthy Dutch food (Bijman et al., 2017), schools, employees of the Federal Office for migration and refugees (in the Netherlands, COA and in Germany das Bundesamt für Migration und Flüchtlinge), volunteers at refugee centres as well as dentists. Social contact with Dutch people may also promote health of refugee children. Prevalence figures show that the current approach within youth healthcare to prevent weight gain in young children is not effective in children with a Turkish background (Wilde et al., 2009; Dommelen et al., 2015). This is expected to be the same for Syrian children. New preventive interventions, such as Simple-Fit!, which has been developed for parents and children with a Turkish background, are more widely implemented in the Netherlands (www.SimpleFit.nl). This intervention can be translated to the Syrian population, which seems necessary.

15.6 Healthy Children in Sound Communities (Gesunde Kinder in gesunden Kommunen)

From children with a Turkish background, data have been collected from the cross-border study 'Gesunde Kinder in einer Gesunden Kind Umgebung' (GKGK) that was conducted in Germany and the Netherlands from 2009 onward (Naul et al., 2012). In this study, municipality, schools, and sports clubs cooperate, and children aged six participated for three (and sometimes four) years. The aim of the approach is to improve children's physical activity levels to 60–90 minutes a day, offer children education lessons about healthy food and taste at school on a weekly basis, include parents in the program, and promote sleep and rhythm in their daily schedules. Measurements include (1) physical fitness tests (coordination, strength, endurance, flexibility, and speed), (2) lifestyle measurements by questionnaires (physical activity, sport, leisure activity, physical/emotional well-being, support of family and friends) and (3) social benefits measured by questionnaires (physical self-concept, group climate). For an overview of the sample description see tables 15.3-7.

Table 15.3: Amount of Dutch (NL), German (D) and children with a Turkish (TUR) migration background in the GKGK study (*n*=772)

	Turkish	D+ NL	Total
Boys	42	340	382
Girls	41	349	390
Total	83	689	**772**

Table 15.4: Mean age of boys and girls who participated in the GKGK study at t1, t2 and t3

	Age t1	Age t2	Age t3
G+NL boys	7.00	7.94	8.93
G+NL girls	6.94	7.88	8.87
TUR boys	7.13	8.08	9.07
TUR girls	7.01	7.96	8.95
Total	6.98	7.92	8.91

Table 15.5: BMI mean of Dutch, German and Turkish boys and girls at t1 and t3

	Boys D/NL *N*=340	Boys TUR *N*=42	Girls D/NL *N*=349	Girls TUR *N*=41
T1	16.17	17.91	16.14	16.78
T3	17.01	19.27	17.10	17.53
Difference	+0.84 *	+1.36 *	+0.96 *	+0.75 *

Table 15.6: BMI percentiles of Dutch, German and Turkish boys and girls at t1 and t3

	Boys D/NL *N*=340	Boys TUR *N*=42	Girls D/NL *N*=349	Girls TUR *N*=41
T1	4.67	5.57	4.68	5.12
T3	4.60	5.71	4.65	4.88
Difference	−0.07	+0.14	−0.03	**−0.24**

Turkish girls of both countries (DE/NL) have the lowest increase of their BMI mean and the highest reduction of their BMI percentile; German and Dutch boys have a lower BMI mean and a relatively better reduction of their BMI percentile than German and Dutch girls (tables 15.5 and 15.6).

The prevalence of overweight reduced in Dutch and Turkish boys and girls together from approximately 30 percent to 25 percent. On the contrary, prevalence of overweight in German Turkish boys and girls increased from approximately 18 percent to 29 percent. The prevalence of obesity remained stable in German and Dutch Turkish boys and girls together (tables 15.5 and 15.6). Figures 15.1 and 15.2 show the development of overweight and obesity in Dutch, German, and Turkish boys and girls. Turkish Girls – especially in the Netherlands – reduce their rate of overweight and BMI-reference mean significantly.

Table 15.7: Overweight percentages of German and Dutch children and children with a Turkish background in the Netherlands and Germany

Overweight (%)	D all *N*=438 (%)	D Turk *N*=43 (%)	NL all *N*=249 (%)	NL Turk *N*=40 (%)
T1	12.8	4.7	8	22.5
T3	15.3	14.0	7.2	17.5
Difference	+ 2.5	+ 9.3	−0.8%	−5.0

Table 15.8: Obesity percentages of German and Dutch children and children with a Turkish background in the Netherlands and Germany

Obesity (%)	D all *N*= 438 (%)	D Turk *N*=43 (%)	NL all *N*= 249 (%)	NL Turk *N*=40 (%)
T1	**6.4**	**14.0**	**4.0**	7.5
T3	7.3	14.0	3.2	7.5
Difference	+ 0.9	0.0	−0.8	0.0

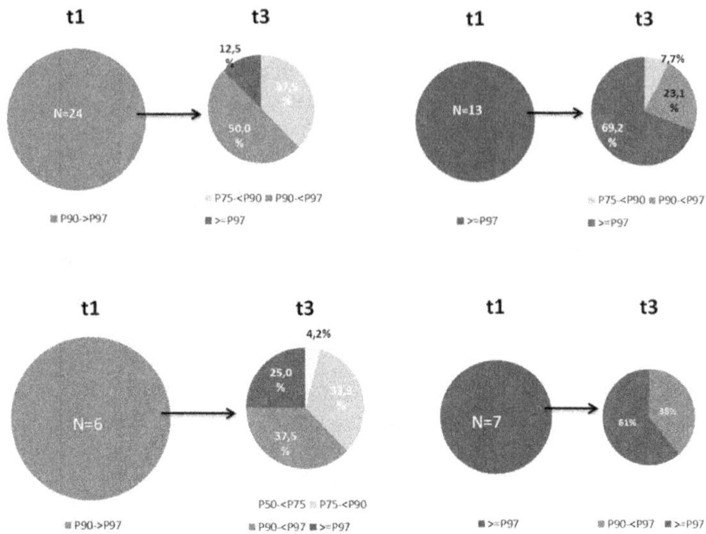

Figure 15.1: Upper part: *longitudinal development of BMI for German and Dutch girls* (red circle = *overweight;* brown circle = *obesity (t1 to t3)* Lower part: *longitudinal development of BMI for Turkish girls* (red circle = *overweight;* brown circle = *obesity (t1 to t3)*

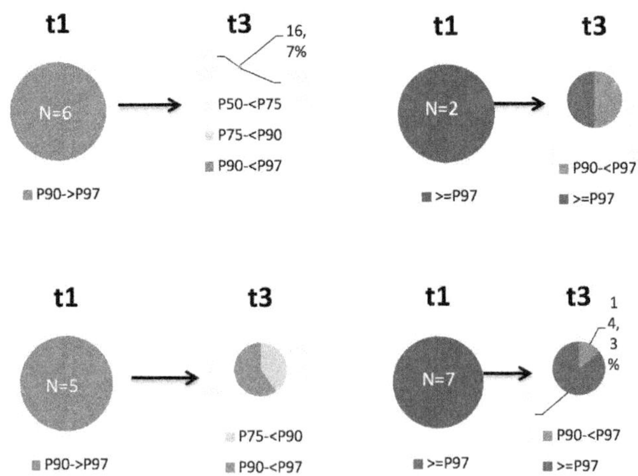

Figure 15.2: Upper part: longitudinal development of BMI for German and Dutch boys (red circle = *overweight;* brown circle = *obesity (t1 to t3)* Lower part: *longitudinal development of BMI for Turkish boys* (red circle = *overweight;* brown circle = *obesity (t1 to t3)*

Furthermore, the GKGK results show that the motor abilities of Turkish children are lower at the start of the project compared to Dutch and German children (fig. 15.3). Turkish children as well as German and Dutch children have a significant progression in motor development, except for the six-minute run (data not shown).

Measurement with questionnaires show that Turkish children not only experience more physical activity, but also spend more time using media than their non-Turkish classmates (fig. 15.4). Finally, especially Turkish girls are well integrated in their classes (fig. 15.5). Similar results are found for media use and physical activity between the sample groups (fig. 15.6) Finally, in Turkish boys very strong correlations are found between their motor abilities and physical self-concept (table 15.7). This is not found in Turkish girls (table 15.8), nor in Dutch and German boys and girls (table 15.9-10).

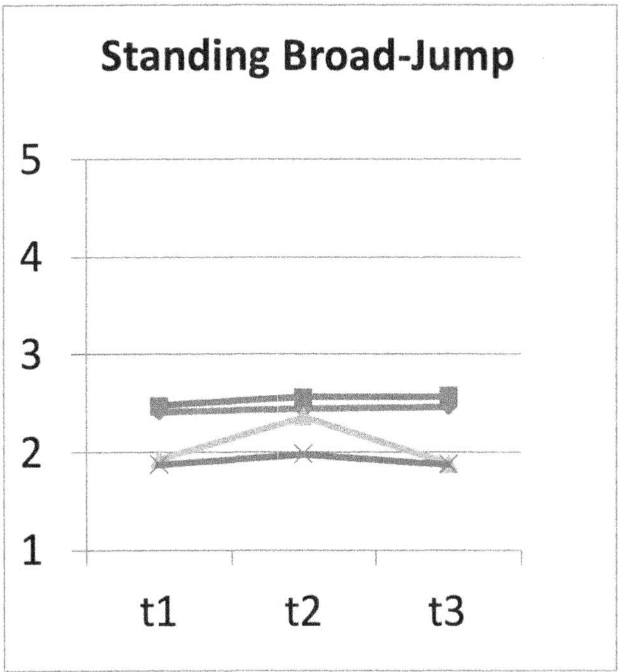

Figure 15.3: Results of motor development (reference norms), SB-jump: Country $p < .001$

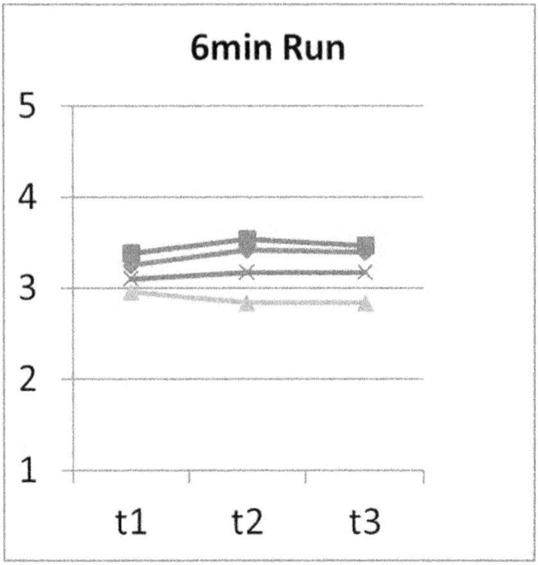

Figure 15.4: Results of motor development (reference norms), 6 min run: Country × time $p < .001$

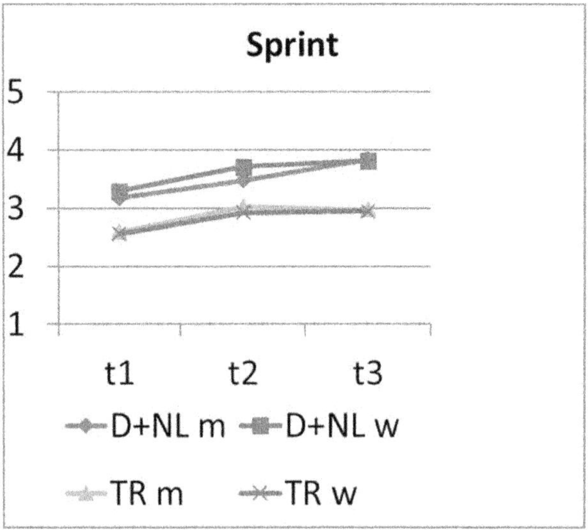

Figure 15.5: Results of motor development (reference norms) sprint: time $p < .001$, *country* $p < .001$

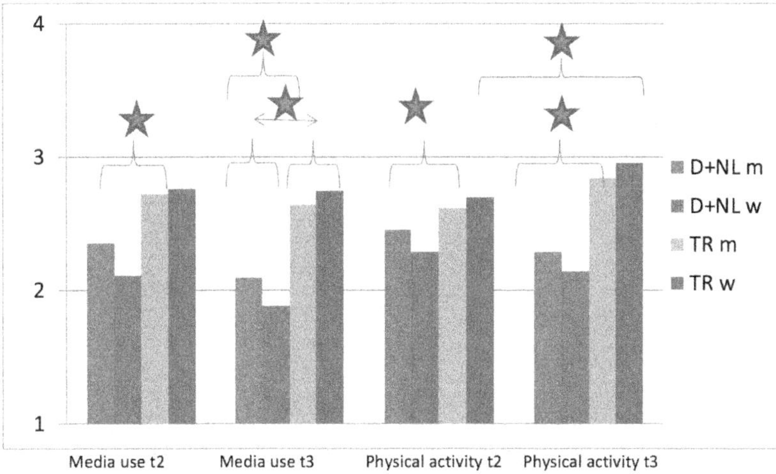

Figure 15.6: Perceived media use at t2 and t3 and perceived physical activity at t2 and t3 of German, Dutch and Turkish boys (m) and girls (w).
Significant higher media and physical activity (PA) behaviour experienced by Turkish children.
Significant positive development of PA behaviour in Turkish children.
Significant Interactions Gender × Country × Activity: Turkish girls (f) rate themselves as more active than boys).

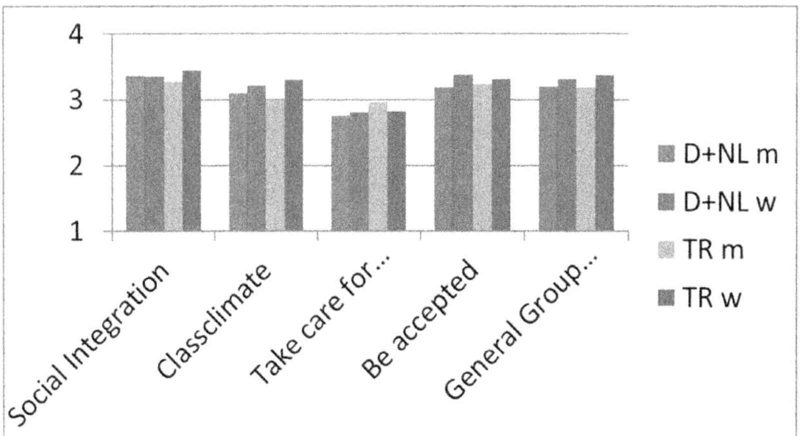

Figure 15.7: Class climate (p = .001), taking care (p = .017), being accepted (p < .001), general group cohesion (p = .001) of German, Dutch and Turkish boys (m) and girls (w)

Table 15.9: Correlations between physical Self-concept and motor development in Turkish boys							
Turkish boys	PSC esport	PSC attractivity	PSC endurance	PSC flexibility	PSC coordination	PSC power	PSC velocity
BMI	−.259	−.155	−.465	−.465	−0,112	.02	−.565
Sit ups (power)						−.036	.258
Push Ups (power)						.201	.222
Standing broad jump (power, coord.)	.313		.364	.445		.222	.449
Jumping sidewards (coord.)			.319		0,038		
Balancing backwards (coord.)					0,019		.317
6 min run (endurance)	.203		.288	.497			.457
20 m run (velocity)	.335		.297	.337			.54

Table 15.10: Correlations between physical self-concept and motor development in Turkish girls							
Turkish girls	PSC sport	PSC attractivity	PSC endurance	PSC flexibility	PSC coordination	PSC power	PSC velocity
BMI	−.078	.095	−.131	−.032	−.18	.189	−.183
Sit ups (power)					−.079	−.225	
Push ups (power)					−.034	−.338	
Standing broad jump (power, coord.)	−.316			−.271		−.283	

(Continued)

Table 15.10: Correlations between physical self-concept and motor development in Turkish girls

Turkish girls	PSC sport	PSC attractivity	PSC endurance	PSC flexibility	PSC coordination	PSC power	PSC velocity
Jumping sidewards (coord.)						−.213	
Balancing backwards (coord.)							
6 min run (endurance)		−.234	0,168			−.26	
20 m run (velocity)							.23

15.7 Discussion

In the longitudinal study "Healthy Children in a Sound Community" (GKGK) in which children at elementary schools are included from six to ten years old (689 German and Dutch children and 83 children with a Turkish background), it is demonstrated that by daily physical activity and weekly lessons on healthy food at school, after three years of intervention, definitely girls with a Turkish background seem to profit most from the intervention. They reduced their BMI, increased their weekly time of physical activity, improved their motor abilities – especially aerobic endurance capacity – although they had high screen/media time levels. Furthermore, they were well integrated in their classes and had a stable, positive self-concept. Turkish boys seem to have a separate position in the family. They are allowed to choose their own behaviour, which in this young age group is not always the desirable behaviour. Such a permissive child-rearing style for boys does not help them to remain healthy and motoric fit. Parents with a Turkish background, as well as those with a Syrian background, should receive support to develop more effective parenting skills when they live in countries like Germany and the Netherlands. Healthcare workers who understand the culture of excessiveness and indulgence, especially in families with Turkish boys, should be involved in preventive programs. Experiences with Simpel-Fit! show that change into a healthy lifestyle for boys and girls of families with a Turkish background is feasible. We expect similar effects in Syrian families. Intervention programs should start early after refugees arrive, because these vulnerable families should receive effective support that helps children to adapt and develop a healthy lifestyle in Germany and the Netherlands.

References

Beenackers, M. A., Nusselder, W. J., Oude, G. J., & Van Lenthe, F. J. (2015). *Het terugdringen van gezondheidsachterstanden: een systematisch overzicht van kansrijke en effectieve interventies. Rapport, Afdeling Maatschappelijke Gezondheidzorg.* Rotterdam: Erasmus MC Universitair Medisch Centrum Rotterdam.

Bijman, C., Nuraydin, S., & L'Hoir, M. (2012). *Voor de nieuwkomers van baby's peuters en kleuters.* Informatie over voeding en gezondheid voor ouders. (In ontwikkeling, 2017).

Bocca, G,. Corpeleijn, E., Stolk, R. P., & Sauer, P. J. J. (2011). Results of a multidisciplinary treatment program in 3-yearold to 5-year-old overweight or obese children a randomized controlled clinical trial. *Archives of Pediatrics & Adolescent Medicine, 166*, 1109–1115.

Boere-Boonekamp, M., L'Hoir, M. P., Beltman, M., Bruil, J., & Dijkstra, N. (2008). *Primaire preventie van overgewicht en obesitas in de JGZ voor 0-4 jarigen, over inpassing van de gedragsdoelen in de JGZ-contactmomenten in Ned.* Tijdschrift Geneel. skd.;152, 324–330.

Bukman, A. J., Teuscher, D. B., Meftah, J., Groenenberg, I., Crone, M. R., van Dijk, S., Bos, M. B., & Feskens, E. J. (2016). *Exploring strategies to reach individuals of Turkish and Moroccan origin for health checks and lifestyle advice. BMC Family Practice, 17*, 85. doi:10.1186/s12875-016-0476-1

De Hoog, M. (2012). *Ethnic differences in metabolic and cardiovascular function in 5 year old children, the role of early life factors.* Amsterdam: Free University.

De Onis, M., & Blossner, M. (2000). Prevalence and trends of overweight among preschool children in developing countries. *American Journal of Clinical Nutrition, 72*(4), 1032–1039.

Van Dommelen, P., Schönbeck, Y., van Buuren, S., & HiraSing, R. A. (2014). Trends in a life threatening condition: Morbid obesity in Dutch, Turkish and Moroccan children in The Netherlands. *PLoS ONE, 9*(4), e94299. doi:10.1371/journal.pone.0094299

Van Dommelen, P., Schönbeck, Y., HiraSing, R. A., & Van Buuren, S. (2015). *Call for early prevention: Prevalence rates of overweight among Turkish and Moroccan children in the Netherlands.* European Journal of Public Health. http://dx.doi.org/10.1093/eurpub/ckv051

van Eijsden, M., Meijers, C. M., Jansen, J. E., De Kroon, M. L., & Vrijkotte, T. G. (2015). Cultural variation in early feeding pattern and maternal perceptions of infant growth. *British Journal of Nutrition, 114*(3), 481–488.

Factsheet. (Pharos2015). *Syrische vluchtelingen.* Retrieved from www.Pharos.nl/documents/doc/factsheet-syrische-vluchtelingen

Factsheet GGD. (2016). *Over gewicht bij Turkse en Nederlandse jeugd.* Retrieved from www.kvnog.nl/images/redactie/bestanden/Factsheets/

German Interior Ministry. (2015). *"Anhaltend hoher Asyl-Zugang im Oktober 2015".* November 5, 2015.

Hales, C. N., & Barker, D. J. (1992). Type 2 (non-insulin-dependent) diabetes mellitus: The thrifty phenotype hypothesis. Diabetologia, *35*, 595–601.

van Hoek, E. (2015). *Young children and obesity: Development and evaluation of familiy-oriented treatment.* Wageningen: Wageningen University.

Hulsmann, A. R., Senyurek, A., & Oostenbrink, R. (2005). Verschillen in voedingsgewoonten in het eerste levensjaar tussen Turkse en Nederlandse zuigelingen. *Tijdschrift voor kindergeneeskunde, 73*(6), 6–9.

Iglowstein, I., Jenni, O. G., Molinari, L., & Largo, R. H. (2003). Sleep duration from infancy to adolescence: reference values and generational trends. *Pediatrics, 111*(2), 302–307.

de Jong-Lenters, M. (2016). *Just add positivity? Dental caries, obesity and problem behaviour in children: the role of parents and family relations.* PhD thesis, Universiteit van Amsterdam.

Küpers, L. K. (2016). *The first 1000 days and beyond: From early life environment to epigenetics and childhood overweight.* Thesis. Groningen: Rijksuniversiteit Groningen.

Landsmeer, N., Hoir, M. L., Schoenmakers, T., & Pillen, S. (2014). Kind en beeldscherm: een te hecht koppel: ouders vinden het moeilijk om bij te sturen. *Medisch Contact, 21*, 1038–1041.

L'Hoir, M. P., Beltman, M., van Sleuwen, B. E., Engelberts, A. C., & Boere-Boonekamp, M. M. (2008). Kansrijke elementen in de preventie van overgewicht bij jonge kinderen [Favourable elements in the prevention of overweight in young children]. *Tijdschrift voor Kindergeneeskunde, 4*(76), 199–205.

L'Hoir, M. P., Boere-Boonekamp, M. M., Beltman, M., Bruil, J., & Engelberts, A. C. (2006). Preventie van overgewicht bij niet-westerse zuigelingen. *Tijdschrift voor jeugdgezondheidszorg, 38*(5), 91–94.

Monasta, L., Batty, G. D., Cattaneo, A., Lutje, V., Ronfani, L., Van Lenthe, F. J., & Brug, J. (2010). Early-life determinants of overweight and obesity: A review of systematic reviews. *Obesity Reviews, 11*(10), 695–708. doi:10.1111/j.1467-789X.2010.00735.x

Naul, R., Schmelt, D., Dreiskaemper, D., Hoffmann, D., & lHoir, M. (2012). 'Healthy children in sound communities'(HCSC/gkgk)—a Dutch–German community-based network project to counteract obesity and physical inactivity. *Family practice, 29*(suppl_1), i110–i116.

Ng, M., Fleming, T., Robinson, M., Thomson, B., Graetz, N., Margono, C., & Abraham, J. P. (2014). Global, regional, and national prevalence of overweight and obesity in children and adults during 1980–2013: A systematic analysis for the Global Burden of Disease Study 2013. *The Lancet, 384(*9945), 766–781.

NJi. Media? Gewoon opvoeden! Toolbox, 2015. Retrieved from www.Nji.nl/mediaopvoeding

Nuraydin, S. (2016). Een Turks kind ontzeg je niets. Reportage Eetcultuur. Volkskrant, 14 maart 2016.

Oudhof, M., de Wolff, M. S., Ruiter, M. D., Kamphuis, M., L'Hoir, M. P., & Prinsen, B. (2013). *JGZ richtlijn opvoedingsondersteuning voor hulp bij opoedingsvragen en lichte opvoedproblemen.* Nederlands Centrum Jeugdgezondheid (NCJ), Utrecht.

Schönbeck, Y., Talma, H., van Dommelen, P., Bakker, B., Buitendijk, S. E., HiraSing, R. A., et al. (2011). Increase in prevalence of overweight in Dutch children and adolescents: A comparison of nationwide growth studies in 1980, 1997 and 2009. PLoS ONE, *6*(11), e27608. doi:10.1371/journal.pone.0027608

Simpel Fit! Retrieved from https://www.loketgezondleven.nl/interventies/i-database/1401160

Whitaker, R. C., Wright, J. A., Pepe, M. S., Seidel, K. D., & Dietz, W. H. (1997). Predicting obesity in young adulthood from childhood and parental obesity. *New England Journal of Medicine, 337*(13), 869–873.

de Wilde, J. A., van Dommelen, P., Middelkoop, B. J., & Verkerk, P. H. (2009). Trends in overweight and obesity prevalence in Dutch, Turkish, Moroccan and Surinamese South Asian children in the Netherlands. *Archives of disease in childhood.* doi:10.1136/adc.2009.163709. Epub 2009 Jun 24

Willik van der, E. M., Vrijkotte, T. G. M., Altenburg, T. M., Gademan, M. G. J., Kist-van Holten, J. (2015). Exclusively breastfed overweight infants are at the same risk of childhood overweight as formula fed overweight infants. Archives of Disease Children, online first, June 4. http://dx.doi.org/10.1136/archdischild-2015-308386

Wit, J. M. (2000). Implicaties van de Barker-hypothese voor de medicus practicus. *Ned Tijdschr Geneeskd, 44*(52), 2491–2494.

www.COA.nl.

Yavuz, H. M., Ijzendoorn van, M. H. Mesman, J., & Veel van der, S. (2015). Interventions aimed at reducing obesity in early childhood: A meta-analysis of programs that involve parents. *Journal of Child Psychology and Psychiatry, 56*(6), 677.

16 Academic Publications of Roland Naul, 1972–2018

Jacob Kornbeck and Dennis Dreiskämper

Von 1972 bis 1979

Naul, R., & Voigt, H. (1972). Aggressionsverhalten bei Volleyballspielern. *Die Leibeserziehung, 21*, 83–87.

Rettberg, U., Naul, R., & Voigt, H. (1972). Erfolg und Zusammenhalt in Sportgruppen. In *Leistungssport*, 2, 128–134. (Übersetzung aus dem Niederlandischen von Koos Vos & Wim Brinkman "Sucess en cohesie in sportgroepenÐ).

Naul, R., & Voigt, H. (1972). Zur Problematik des Divergenztheorems für leistungsorientierte Ballspielmannschaften. *Sportwissenschaft*, 3, 300–306.

Naul, R. (1973). *Probleme fachdidaktisch orientierter Curriculumsforschung –dargestellt am Beispiel des Unterrichtskomplexes Arbeitslehre.* Münster: Westfälische Wilhelms Universität Münster (Magisterarbeit).

Voigt, H., & Naul, R. (1973). Zur Diagnostik der Aggressivität im Sport. In O. Gruppe & H. Baitsch (Eds.), *Sport in unserer Welt - Chancen und Probleme. Olympischer Kongreß München vom 21–25 August 1972* (pp. 426–427). Berlin: Springer.

Naul, R., & Voigt, H. (1974). Aggression und Sport. *Soziale Welt*, 25, 347–369.

Naul, R. (1976). Alte Schule mit neuer Legitimation. *Zeitschrift für Pädagogik*, 22, 451–457.

Naul, R. (1976). Besprechung der Unterrichtsmaterialien für den Sportunterricht in der gymnasialen Oberstufe: H. Zimmermann, Aggression und Sport. *Sportunterricht*, 26, 22–24.

Naul, R. (1976). Sport als Lernfeld in der Kollegschule NW. In Ausschuss Deutscher Leibeserzieher (Ed.), *Sport – Lehren und Lernen, Sport - Lehren und Lernen* (pp. 291–293). Schorndorf: Hofmann.

Naul, R., & Blankertz, H. (1976). *Strukturvorgabe für den Schwerpunkt 15 „Erziehung und Soziales"*. Münster/Düsseldorf: WBK.

Naul, R. (1977). Das Normenbuch Sport und seine Folgen für die Reform der Sekundarstufe II. *Sportunterricht*, 26, 184–189.

Fingerle, K. H., Naul, R., & Schenk, B. (1978). Schwerpunktbezogene Grundbildung in der Kollegschule: Problemstellungen und Konstruktionswege – Bilanz und Perspektiven. In B. Schenk & A. Kell (Eds.), *Grundbildung: Schwerpunktbezogene Vorbereitung auf Studium und Beruf in der Kollegschule*: (pp. 255–284). Königstein/Ts.: Athenäum.

Naul, R., Heursen, G., & Gruschka, A. (1978). Die sozialberufliche Grundbildung – Konzept und Struktur. In B. Schenk & A. Kell (Eds.), *Grundbildung: Schwerpunktbezogene Vorbereitung auf Studium und Beruf in der Kollegschule* (pp. 195–221). Königstein/Ts.:Athenäum.

Naul, R. (1978). *Analyse, Konstruktion und Evaluation eines doppelqualifizierenden Schwerpunktprofils in der Kollegschule, dargestellt am Beispiel eines Sportprofils im Schwerpunkt "Erziehung und SozialesĐ.* Münster: Westfälische Wilhelms-Universität (Dissertation zum Dr. phil.).

Naul, R., & Gruschka, A. (1978). *Evaluationsstudie "Sozialberufliche GrundbildungĐ.* Münster: WBK.

Naul, R., & Hiersemann, D. (1978). *Profilkonzept für das doppelt qualifizierende Schwerpunktprofil „Freizeitsportleiter/Allgemeine Hochschulreife"*. Düsseldorf: KOSEK.

Von 1980 bis 1989

Naul, R. (1980). Fünf Jahre Normenbuch Sport – nur ein Sturm im Wasserglas? *Sportpädagogik*, 4(6), 8–9.

Naul, R., Bieler, L., Böhner, D., Hellmich, A., Klaeren H., Kuretzky, R., Siegel, J., & Tücking, H. (1980). Verbindung zwischen Sportpraxis und Sporttheorie im Leistungsfach Sport der Kollegschule NW. In Ausschuss Deutscher Leibeserzieher (Ed.), *Theorie in der Sportpraxis* (pp. 115–124). Schorndorf: Hofmann.

Naul, R., & Holzhausen, H. (1980). *Sport – Eva 1.* Münster: WBK. (1. Zwischenbericht zur Eingangsbefragung der Evaluationsstudie).

Voigt, H., & Naul, R. (1980). *Volleyball: Sport in der Sekundarstufe II*. Düsseldorf: Bagel. Insgesamt 8 Auflagen und 6 Nachdrucke der 8. Auflage (später Schwann, Cornelsen).

Voigt, H., & Naul, R. (1980). *Arbeitsblätter für Volleyballspieler. Angriffschlag*. Düsseldorf: Schwann (2. Auflage 1987).

Voigt, H., & Naul, R. (1980). *Arbeitsblätter für Volleyballspieler. Aufschlag von unten*. Düsseldorf: Schwann (2. Auflage 1987).

Voigt, H., & Naul, R. (1980). *Arbeitsblätter für Volleyballspieler. Frontaler Angriffsschlag*. Düsseldorf: Schwann (2. Auflage 1987).

Voigt, H., & Naul, R. (1980). *Arbeitsblätter für Volleyballspieler. Beidarmige Abwehr im Stand*. Düsseldorf: Schwann (2. Auflage 1987).

Voigt, H., & Naul, R. (1980). *Arbeitsblätter für Volleyballspieler. Einerblock*. Düsseldorf: Schwann (2. Auflage 1987).

Voigt, H., & Naul, R. (1980). *Arbeitsblätter für Volleyballspieler. Oberes Zuspiel*. Düsseldorf: Schwann (2. Auflage 1987).

Voigt, H., & Naul, R. (1980). *Arbeitsblätter für Volleyballspieler. Unteres Zuspiel*. Düsseldorf: Schwann (2. Auflage 1987).

Naul, R. (1982). Theorie – Praxis – Probleme im Sport. In D. Quanz & F. Weckenbrock (Eds.), *Handreichung für den Schulsport 2* (pp. 7–26). Neuss: Landesinstitut für Curriculumentwicklung, Lehrerfortbildung und Weiterbildung.

Naul, R., & Holzhausen, H. E. A. (1982). *Sport – Eva 2: 2. Zwischenbericht zur Eingangsbefragung der Evaluationsstudie*. Münster: WBK.

Holzhausen, H., & Naul, R. (1983). *Absolventenbefragung der Doppelqualifikation Sport, 1978–1981 (Eva 3)*. Münster: WBK.

Naul, R. (1983). Analysen zum Unterricht im Leistungsfach Sport aus Schülersicht. *Leibesübungen - Leibeserziehung*, 37, 195–200.

Naul, R. (1983). Unterricht: Sport. In D. Lenzen (Ed.), Enzyklopädie Erziehungswissenschaft: Sekundarstufe II, *Jugendbildung zwischen Schule und Beruf* (Band 9.2, pp. 645–652). Stuttgart: Klett-Cotta.

Naul, R. (1983). Wahlbereich. In D. Lenzen (Ed.), Enzyklopädie Erziehungswissenschaft: Sekundarstufe II, *Jugenbildung zwischen Schule und Beruf* (Band 9.2. pp. 994-996). Stuttgart: Klett-Cotta.

Naul, R. (1984). *Die Struktur und Entwicklung sportspezifischer Kompetenzen beiGymnasiasten und Kollegschülern*: (Kurzfassung). Essen: UGE.

Naul, R., & Fischer, B. (1984). Sportliche Sozialisation, Einstellung und Kompetenzentwicklung bei Schülern der Sekundarstufe II mit dem Leistungsfach Sport. In H. Rieder (Ed.), *Motorik und Bewegungs-Forschung im Sport* (Band 2, pp. 152–175). Köln: Strauß.

Walther, K.-D., & Naul, R. (1984). *Sport – Eva 6: Die Struktur und Entwicklung der Rollen-Kompetenz bei Kollegschülern und Gymnasiasten.* Essen: Universität Erfurt (Verlag).

Naul, R. (Ed.). (1985). *Körperlichkeit und Schulturnen im Kaiserreich.* Wuppertal: Putty (Puttys Sport-Paperback Reihe Theorie Band 2).

Naul, R. (1985). Fallstudie zur Sozialgeschichte des Turnunterrichts - das Helmholtz- Realgymnasium in Essen. In R. Naul (Ed.), *Körperlichkeit und Schulturnen im Kaiserreich* (pp. 95–142). Wuppertal: Putty.

Naul, R., & Breyvogel, W. (1985). Das Gymnasium und sein Berechtigungsmonopol. In R. Naul (Ed.), *Körperlichkeit und Schulturnen im Kaiserreich* (pp. 65–92). Wuppertal: Putty

Naul, R. (1985). Sport in der Schule. In W. Twellmann (Ed.), *Handbuch Schule und Unterricht* (Band 7.2, pp. 751–776). Düsseldorf: Schwann.

Naul, R. (1985). The development of physical education in German secondary schools during the period of the industrial revolution, 1870–1915. In N. Müller & J. K. Rühl (Eds.), *Olympic Scientific Congress sports history 1984. Official Report. Sport history* (pp. 457–467). Niedernhausen: Schors.

Naul, R., & Gruschka, A. (1985). Empirische Studien zur Kollegschule. In A. Gruschka (Ed.), *Wie Schüler Erzieher werden: Studie zur Kompetenzentwicklung und der fachlichen Identitätsbildung in einem doppelt-qualifizierten Bildungsgang des Kollegschulversuchs NW, Wie Schüler Erzieher werden* (pp. 10–14). Wetzlar: Büchse der Pandora.

Naul, R. (1986). Fußball in den USA. *Sportpädagogik, 10*(5), 7–8.

Naul, R. (1986). Fußballspielen in der Schule. *Sportpädagogik, 10*(5), 14–25.

Naul, R. (1986). Schulsportgeschichte. *Sportunterricht*, 35, 445–447.

Naul, R. (1986). Sportstudie. In H. Blankertz (Ed.), *Lernen und Kompetenzentwicklung in der Sekundarstufe II: Curriculumentwicklung in Nordrhein-Westfalen* (pp. 88–96, 130–137, 569–651). Soest: LSW.

Naul, R., Fischer, B., Schenk, B., Gruschka, A., & Meyer, M. A. (1986). Schüler auf dem Weg zu Studium und Beruf. Zeitschrift für Pädagogik, 32, 557–577.

Naul, R., & Kuretzky, R. (1986). Fußball als Kultur und Kommerz. *Sportpädagogik, 10*(5), 55–60.

Allkemper, G., & Naul, R. (1987). *Spiele im Haus: Vom Raten, Wetten und Tanzen.* Münster: Coppenrath.

Allkemper, G., & Naul, R. (1987). *Spiele in Eis + Schnee: Von Iglus, Schneemännern und Rodelschlitten.* Münster: Coppenrath.

Brauweiler, F., Naul, R., & Schippert, D. (1987). *Berufsschulsport: Aufgaben und Chancen für die Unterrichtspraxis (Band 53).* Bremen: Wissenschaftliches Institut für Schulpraxis.

Naul, R. (1987). Die Diskussion über die körperliche Erziehung auf der Reichschulkonferenz 1920. In R. Geßmann (Ed.), *Schulturnen zur Zeit der Weimarer Republik, Schulturnen zur Zeit der Weimarer Republik* (pp. 46–69). Köln: Strauß.

Naul, R. (1987). Sporterziehunq als Bestandteil einer neuen Allgemeinbildung. In H. Heid & H. G. Herrlitz (Eds.), *Allgemeinbildung. 10. DGfE-Kongreß. 21. Beiheft der Zeitschrift für Pädagogik* (pp. 161–171). Weinheim: Beltz.

Naul, R. (1987). *Straßenakrobaten, Bewegungsroboter und Traumtänzer. Sportpädagogik, 11*(3), 6–7.

Naul, R. (1987). Turnen nach Jahn oder Spieß? In J. A. Mangan (Ed.), *Proceedings of the XI. HISPA International Congress 1985, Glasgow, Proceedings of the XI. HISPA International Congress 1985, Glasgow* (pp. 235–238). Glasgow: Univ.

Naul, R., Fischer, B., & Falkenberg, G. (1987). Auswertung kognitiven Lernens um Leistungsfach Sport der gymnasialen Oberstufe. In Ausschuss Deutscher Leibeserzieher (Ed.), *Sport – Planen, Durchführen, Auswerten* (pp. 189–191). Schorndorf: Hofmann.

Naul, R., & Kuretzky, R. U. A. (1987). *Freizeitsportleiter – Bildungsgangbeschreibung* (Reihe Kollegschule). Soest: LSW.

Naul, R., Tiegel, G., & Schmidt, W. (Eds.). (1987). *Beiträge und Analysen zum Fußballsport,* I. Clausthal-Zellerfeld: Greinert.

Naul, R., Schmagold, J., & Wadle, H.-J. (1987). Zum Studium des Faches Fußball an bundesdeutschen Hochschulen im Rahmen der Sportlehrerausbildung. In R. Naul, G. Tiegel, &

W. Schmidt (Eds.), *Beiträge und Analysen zum Fußballsport* (pp. 153–162). Clausthal-Zellerfeld: Greinert.

Allkemper, G., & Naul, R. (1988). *Spiel dich gesund: Spiele für Kinder rund ums Krankenbett.* Münster: Coppenrath.

John, H.-G., & Naul, R. (Eds.). (1988). *Jugendsport im ersten Drittel des 20. Jahrhunderts.* Clausthal-Zellerfeld: Greinert.

Naul, R. (1988). *Besprechung der Festschrift für H. Bernett zum 65. Geburtstag. Sportwissenschaft,* 18, 75–80.

Naul, R. (1988). Female soccer coaches' socialization, attitudes towards soccer, and their assessment of training as licensed coaches. In T. Reilly (Ed.), *Science and football. proceedings of the first congress of science and football* (pp. 630–636). London/New York: E. & F.N. Spon.

Naul, R. (1988). Learning about Sports: A cognitive approach and its evaluation in the German Grammar School physical education curriculum. *International Journal of Physical Education,* 25, 16–22.

Naul, R. (1988). Von Potsdam nach Weimar. In H.-G. John & R. Naul (Eds.), *Jugendsport im ersten Drittel des 20. Jahrhunderts* (pp. 84–130). Clausthal-Zellerfeld: Greinert.

Naul, R. (1989). *Berufsanalyse für Spitzensportler und Netzplanung für flexible Qualifikationswege in der beruflichen Laufbahn der Athleten: (Forschungsbericht für die Deutsche Sporthilfe).* Essen: UGE.

Naul, R. (1989). Besprechung der Bücher Schultz,U. (Hg.), Das Fest, München 1988 (Beck) und Düding, Friedmann, Münch (Hg.) Öffentliche Festkultur im 19. Jahrhundert, Hamburg 1988 (Rowohlt). *Stadion,* 15, 294–300.

Naul, R. (1989). Gymnastics, athletics, games: sedan celebrations in imperial Germany. In M. Lämmer (Ed.), *Proceedings of the XII HISPA Congress* (pp. 131–136). St. Augustin: Academia.

Naul, R., & Großbröhmer, R. (1989). Carl Euler – Pionier der Turnlehrerbildung und Historiker des Schulturnens in Preußen. In M. Lämmer (Ed.), *Proceedings of the XII HISPA Congress* (pp. 265–272). St. Augustin: Academia.

Naul, R. (1989). Olympic headquarters: a new challenge to sport pedagogy in the Federal Republic of Germany. In F. Fu, M. Ng, & M. Speak. (Eds.), *Comparative physical education and sport* vol. 6 (pp. 199–204). Hong Kong: Chinese University Press.

Naul, R., & Schmidt, W. (Eds.). (1989). *Beiträge und Analysen zum Fußballsport 2: Referate zu den Tagungen der dvs-Kommission Fussball 1987 und 1988 in Duisburg*. Clausthal-Zellerfeld: Greinert.

Naul, R. (1989). Sportwissenschaftliche Analysen zum Frauenfußball. In R. Naul & W. Schmidt (Eds.), *Beiträge und Analysen zum Fußballsport 2* (pp. 38–60). Clausthal-Zellerfeld: Greinert.

Naul, R. (1989). The Olympic athlete career: The role of social counseling and vocational training. In United States Olympic Committee (Ed.), *First IOC Congress on sport sciences* (pp. 414–415). Colorado Springs, CO: USOC.

Naul, R. (1989). The rise and fall of the German national olympic games. In L. Zecevic (Ed.), ICOSH- Seminar 1988. The Olympic Movement "Past-present and future" (pp. 222–227). Sarajevo: ICOSH.

Naul, R. (1989). Turnen, Spiel und Sport der Schuljugend im ersten Drittel unseres Jahrhunderts. In W.-D. Brettschneider, J. Baur, & M. Bräutigam (Eds.), *Bewegungswelt von Kindern und Jugendlichen. Bericht über den 8. Sportwissenschaftlichen Hochschultag der Deutschen Vereinigung für Sportwissenschaft, Paderborn 1987* (pp. 338–349). Schorndorf: Hofmann.

Naul, R., & Winkelmann, J. (1989). 125 Jahre Turnen, Spiel und Sport am Helmholtz-Gymnasium. In U. Althaus (Ed.), *Helmholtz-Gymnasium Essen-125 Jahre Schule* (pp. 93–99). Essen: Schulleitung und Schulgemeinde des Helmholtz-Gymnasiums Essen.

Von 1990 bis 1999

Naul, R. (1990). Pädagogische Probleme im Kinder- und Jugendtraining. *Wissenschaftliche Zeitschrift der DHfK Leipzig*, 31, 447–452.

Naul, R. (1990). Teaching P.E. in Germany – An historical view. In R. Telama (Ed.), *Physical education and life-long physical activity* (pp. 342–349). Jyväskylä: AEISEP World Congress.

Naul, R. (1990). The renaissance of the history of school sports: Back to the future? *Journal of Sport History*, 17, 199–213.

Naul, R. (1990). Volleyball: wissenschaftpropädeutisch lernen. *Sportpädagogik*, 14(2), 76–80.

Naul, R., & Rodermund, J. (1990). Arbeitsbeschaffungsmassnahmen für Sportlehrer. *Spectrum der Sportwissenschaft*, 2, 105–119.

Helmke, C., Naul, R., & Rhode, J. (1991). Zur Lehrplanentwicklung und Lehrplanreform des Sportunterrichts in der ehemaligen DDR und in den neuen Bundesländern. *Sportunterricht*, 40, 382–394.

Ichimura, S., & Naul, R. (1991). Cross-Cultural assessments and attributions to female soccer: Japan and West Germany. In J. Standeven, K. Hardman, & D. Fisher (Eds.), *Sport for all. Into the 90s. Comparative physical education and sport* (pp. 212–220). Aachen: Meyer & Meyer.

Krotee, M., Huang, C., & Naul, R. (1991). A cross-cultural study of the socialisation process of female soccer athletes. In J. Standeven, K. Hardman, & D. Fisher (Eds.), *Sport for all. Into the 90s; Comparative physical education and sport.* (pp. 242–248). Aachen: Meyer & Meyer.

Naul, R., Pauer, M., & Rychtecky, A. (1991). Daily physical activities and motor performance of West German and Czechoslovakian School Children. In J. Standeven, K. Hardman, & D. Fisher (Eds.), *Sport for all. Into the 90s. Comparative physical education and sport* (pp. 204–211). Aachen: Meyer & Meyer.

Naul, R., & Quanz, R. (1991). Präventive Sportpädagogik und Laufbahnberatung. In M. Bührle & M. Schurr (Eds.), *Leistungssport. Herausforderung für die Sportwissenschaft; Bericht über den 9. Sportwissenschaftlichen Hochschultag der Deutschen Vereinigung für Sportwissenschaft, Freiburg 1989* (p. 237). Schorndorf: Hofmann.

Naul, R. (1991). Aufgaben präventiver Sportpädagogik: Sozial-ökologische Karriereforschung im Leistungssport. In M. Bührle & M. Schurr (Eds.), *Leistungssport. Herausforderung für die Sportwissenschaft; Bericht über den 9. Sportwissenschaftlichen Hochschultag der Deutschen Vereinigung für Sportwissenschaft*,(pp. 238–241). Schorndorf: Hofmann.

Naul, R. (1991). Die Berufsanalyse für Spitzensportler als Teil einer sozialökologischen Karriereforschung. In M. Bührle & M. Schurr (Eds.), *Leistungssport. Herausforderung für die Sportwissenschaft; Bericht über den 9. Sportwissenschaftlichen Hochschultag der Deutschen Vereinigung für Sportwissenschaft* (pp. 243–246). Schorndorf: Hofmann.

Naul, R. (1991). Female soccer – A different game to play? *Science and Soccer*, 4, 24–30.

Naul, R. (1991). *Freizeit und Wirtschaft im Jahr 2000 – Ein Nachwort (Freizeit und Wirtschaft im Jahr 2000, Ergebnisse des 1. Inpro-Symposiums*). Essen: INPRO.

Naul, R. (1991). Learning about sports. A cognitive approach and its evaluation in the German grammar school P.E. Curriculum. In H. Barreiros (Ed.), *Actes de la Conférence Mondiale Motricité Humaine* (pp. 131–138). Lisboa: FMH.

Naul, R. (Ed.). (1991). *Turnen and sport: The cross-cultural exchange*. Münster: Waxmann. (German and American Studies in Sport, vol. 1).

Naul, R. (1991). Studies of the turner system by an American: Fred E. Leonard`s Visits to Germany, 1896, 1900/01, 1913. In R. Naul (Ed.), *Turnen and sport. The cross-cultural exchange*, (pp. 121–145). Münster: Waxmann.

Naul, R., & Jonischeit, L. (1991). *Jugendsport im deutschen Kaiserreich und in der Weimarer Republik*. Essen: WGI (Abschlußbericht für die Alfried Krupp von Bohlen und Halbach Stiftung).

Naul, R., & Jonischeit, L. (1991). Spuren des Natürlichen Turnens in Deutschland. In S. Grössing (Ed.), *Margarete Streicher – Ein Leben für die Leibeserziehung. Schriftenreihe des Streicher-Archivs* (pp. 46–85). Salzburg: Streicher Archiv.

Naul, R., & Rodermund, J. (1991). Sportlehrer im ausserschulischen Berufsfeld Sport: Zum Verlauf von Berufskarrieren in Arbeitsbeschaffungsmassnahmen. *Sportunterricht*, 40, 302–308.

Naul, R., & Rodermund, J. (1991). Unemployed P.E. teachers – Their life cycle and job placement after graduation. *AIESEP Newsletter*, 34, 3–5.

Naul, R. (1991). How can sport pedagogy move towards excellence? In T. Williams, L. Almond, & Sparkes A. (Eds.), *Sport and physical activity: Moving towards excellence*. (pp. 52-60). Proceedings of the AIESEP World Convention, July 20–25, 1990. London: E.&F.N.Spon.

Naul, R. (1992). Die "organisierte Freizeit": Vereine und Verbände. In INPRO (Ed.), *Freizeitinfrastruktur in Städten. (2. Inpro-Symposium, Beiträge zur Freizeit-Forschung)* (pp. 76–95). Essen: INPRO.

Naul, R. (1992). Elite sport in Germany: The 1990s. *Journal of the International Council for Health, Physical Education and Recreation, 28*(2), 17–22.

Naul, R. (1992). Fußballsport. Von Fußpanzern, Bajazzo-Hemden und langen Hosen, die immer kürzer wurden. In Deutsches Textilmuseum (Ed.), *Ausstellungskatalog "Sportswear"* (pp. 56–59). Krefeld: Joh. van Aacken.

Naul, R. (1992). Zur Geschichte der Schulturnkleidung. In Deutsches Textilmuseum (Ed.), Ausstellungskatalog "Sportswear" (pp. 18–23). Krefeld: Joh. Van Aacken.

Naul, R. (1992). German unification: Curriculum development and physical education at school in East Germany. *British Journal of Physical Education, 23*(4), 14–19.

Naul, R. (1992). Teaching physical education teachers in the Federal Republic of Germany: A Collaborative Enterprise. In G. M. Graham & M. A. Jones (Eds.), *Collaboration between researchers and practitioners in physical education: An international dialogue* (pp. 127–132). Proceedings from the AIESEP-NAPEHE World Congress. Reston: AAHPERD.

Naul, R., & Blair, P. (1992). A cross-cultural study of youth's attitudes concerning the role of sport in the educational process. *Journal of Comparative Physical Education and Sport,* 14, 4–15.

Naul, R., Falkenberg, G., & Fischer, R. (1992). Kognitives Lernen im Leistungsfach Sport der gymnasialen Oberstufe. *Sportunterricht*, 41, 101–114.

Naul, R., & Hardman, K. (1992). Interschool sports competition: Historical-cultural antecedents in England and Germany. *Journal of Comparative Physical Education and Sport,* 14, 15–30.

Naul, R., & Helmke, C. (1992). Der Oberstufenunterricht in den neuen Bundesländern. *Sportunterricht*, 41, 297–300.

Naul, R., & Schulz, N. (1992). 20 Jahre Leistungsfach Sport – Versuch einer Bilanz. *Sportunterricht*, 41, 274–283.

Naul, R. (1993). Are national Olympic Games pseudo Olympics? Willibald Gebhardt and the German National Festival Movement. In N. B. Bouchier (Ed.), *1993 NASSH Proceedings* (p. 47). Hamilton: NASSH.

Naul, R. (1993). Changes in sport & physical education in East Germany. In R. Fischer (Ed.), *The changing face of physical education & sport in Europe* (pp. 32– 42). Twickenham: St. Mary`s College.

Naul, R. (1993). Court Soccer – eine neue Sportart. *ASB-Zeitung*, 1, 7–8.

Naul, R. (1993). Die neue Sportwissenschaft Ost: Akademischer Offenbarungseid der alten Sportwissenschaft West? *dvs informationen*, 3, 15–19.

Naul, R., Neuhaus, W., & Rychtecky, A. (1993). The significance of daily physical activities for special motor performance: A cross-cultural longitudinal pilot study. In Faculty of Physical Education of Charles University (Ed.), *Proceedings of the 6th ICHPER Europe Congress* (pp. 335–344). Prague: FTVS.

Naul, R. (1994). *Court-Soccer: Regelheft*. Aachen: Meyer & Meyer.

Naul, R. (1994). Das "Internationale Komitee für Sportpädagogik". *dvs informationen*, 2, 24–26.

Naul, R. (1994). Fazit beim DOI-Symposium: Würdigung Gebhardts steht noch aus. In *NOK Report Nr. 12*, 21–23.

Naul, R. (1994). German unification: Decline or regeneration in physical education and sport sciences? In R. C. Wilcox (Ed.), *Sport in the global village* (pp. 281–288). Morgantown, WV: Fitness Information Technology.

Naul, R. (1994). Historical perspectives in sport pedagogy. *Sport Science Review*, 3(1), 11–17.

Naul, R. (1994). Physical education teacher training – Historical perspectives. In J. Mester (Ed.), *Sport, leisure, and physical education. Conference proceedings: Current and future perspectives* (pp. 588–610). Aachen: Meyer & Meyer.

Naul, R., & Breuer, H. (Eds.). (1994). *Schwimmsport und Sportgeschichte. Zwischen Politik und Wissenschaft. Festschrift für Hans-Georg John zum 65. Geburtstag*. Sankt Augustin: Academia.

Naul, R., & Breuer, H. (1994). Hans-Georg John. Hallore, Leibeserzieher und Historiker der Schwimmkunst. In H. Breuer & R. Naul (Eds.), *Schwimmsport und Sportgeschichte: Zwischen Politik und Wissenschaft* (pp. 11-13). *Festschrift für Hans-Georg John zum 65.* Geburtstag. Sankt Augustin: Academia.

Naul, R. (1994). Sind Nationale Olympische Spiele pseudo-olympische Spiele? Willibald Gebhardt und das deutsch-nationale Olympia. In H. Breuer & R. Naul (Eds.), *Schwimmsport und Sportgeschichte: Zwischen Politik und Wissenschaft* (pp. 61–72). *Festschrift für Hans-Georg John zum 65. Geburtstag*. Sankt Augustin: Academia.

Naul, R. (1994). Sportdidaktik nach Mölln – Rückzug in die emotionale Betroffenheit oder Aufbruch zur geistigen Selbstreflexion? *Sportunterricht*, 43, 122–126.

Naul, R. (1994). The elite athlete career: Sport pedagogy must counsel social and professional problems in life development. In D. Hackfort (Ed.), *Psycho-social issues and interventions in elite sports* (pp. 237–258). Frankfurt am Main: Lang. (Sport Sciences International vol. 1).

Hardman, K., & Naul, R. (1994). Attitudes to interscholastic sport competition in England and West Germany: Divergence or convergence? In P. Duffy & L. Dugdale (Eds.), *Health, Physical education and recreation – Moving toward the 21st century* (pp. 105–118). Champaign, IL: Human Kinetics.

Naul, R., & Hardman, K. (1994). The development of sporting excellence in England and Germany. In R. C. Wilcox (Ed.), *Sport in the global village* (pp. 449–458). Morgantown, WV: Fitness Information Technology.

Naul, R., Krotee, M., & Blair, P. (1994). A six-nation study concerning attitudes and participation patterns of youth toward competitive sport. In R. C. Wilcox (Ed.), *Sport in the global village* (pp. 467–475). Morgantown, WV: Fitness Information Technology.

Naul, R., Neuhaus, W., & Rychtecky, A. (1994). Daily physical activities and motor performance of West German and Czechoslovakian schoolchildren. In R. C. Wilcox (Ed.), *Sport in the global village* (pp. 477–486). Morgantown, WV: Fitness Information Technology.

Naul, R., & Krüger, A. (1994). "Geturniert wart den Sumer vil."– Theorie und Praxis der traditionellen Formen aus Leichtathletik und Bewegungsspiel. In U. Becker (Ed.), *Leichtathletik im Lebenslauf* (pp. 395–399). Breitensport- Kongreß des Deutschen Leichtathletik-Verbandes vom 23. bis 25. Oktober 1992 in Mainz. Aachen: Meyer & Meyer.

Naul, R., & Sonnenschein, W. (1994). Internationale sportwissenschaftliche Vereinigungen. *dvs informationen*, 2, 7–8.

Krotee, M. L., Blair, P. F., Hardmann, K., & Naul, R. (1995). Psychosocial patterns of youth toward interscholastic sport: Across-cultural analysis. In B. Svoboda & A. Rychtecky (Eds.), *Physical activity for life: East and West, South and North* (pp. 168–172). Aachen: Meyer & Meyer.

Naul, R., & Pohl, A. (1995). Sportive lifestyle, motor performance, and Olympic ideal of youth in Europe: Preliminary notes on a new comparative youth sport study. In B. Svoboda & A. Rychtecky (Eds.), *Physical activity for life: East and West, South and North* (pp. 209–214). Aachen: Meyer & Meyer.

Svoboda, B., & Naul, R. (1995). Comparison of determinants in motivation of pupils to physical education at secondary schools in Essen (Germany) and in the Czech Republic. In B. Svoboda & A. Rychtecky (Eds.), *Physical activity for life: East and West, South and North* (pp. 309–311). Aachen: Meyer & Meyer.

Naul, R. (1995). Court soccer – Eine motivierende Variante des Fußballspiels. In G. Gerrisch (Ed.), *Psychologie im Fußball* (pp. 147–154). Tagung der dvs-Kommission Fußball vom 17. – 19.11.1993 in Hennef. Clausthal-Zellerfeld: Greinert.

Naul, R. (1995). Bedeutung der Freizeitwirtschaft im Europa der Zukunft. – Herausforderung an Wirtschaft und Wissenschaft. In A.-A. Pourzal (Ed.), *Essen im Dialog. Forumsband zum Wirtschaftsforum 94/95 – aus Anlaß des EU-Gipfels* (pp. 243–345). Essen: FET.

Naul, R. (1995). European studies in Youth's physical activity: Challenges to develop new active lifestyles for children in developing countries? In E. H. Katzenellenbogen (Ed.), *The importance of children`s participation in physical and sporting activities* (pp. 35–51). Stellenbosch: Brand.

Naul, R. (1995). Wie kann man Squash mit entwicklungsgemäßen Spielformen für Kinder und Jugendliche attraktiver machen? In DSP (Ed.), *DSP-Informiert: Aufbruch für die Jugend* (pp. 9–15). Duisburg: Selbstverlag.

Naul, R., & Neuhaus, W. (1995). Actieve levenswijse en lichamelijke fitness bij Duitse jongens en meisjes von 12 tot 14 jaar. *Tijdschrift voor Lichamelijke Opvoeding*, 157(3), 2–7.

Naul, R., Neuhaus, W., & Rychtecky, A. (1995). La Importancia de las Actividades Fisicas Diaturaciones Motrices Especiales: Un Estudio Piloto Inter-Cultural y Longitudinal. In M. Noguera (Ed.), *Actas Congreso Cientifico Olympico 1992. Pedagogica y Edaucation Fisica Comparada Vol. III* (pp. 189–198). Malaga: Instituto Andaluz del Deporte.

Naul, R., Pohl, A., Telama, R., Almond, L., & Pieron, M. (1995). Sportive lifestyle, motor performance, and olympic ideal of youth in Europe: Prelimeinary notes on a new comparative youth sport study. *ICSP*, 10, 2–7.

Naul, R. (1996). Quels Criteres Pour un Degre de "Master en Sciences du Sport" *Bulletin de l` AFRAPAS*, 9, 11–13.

Naul, R. (1996). *Sporting lifestyle, motor performance, and Olympic ideals of youth in Europe: (Summary)*. Essen: WGI.

Naul, R. (1996). *Zukünftige technische Ausstattungs- und Spielmöglichkeiten in Sporthallen* (Vol. 1). Köln: BISP.

Naul, R. (Ed.). (1996). *Zukunftssportarten und ihre Vermarktungspotentiale. Leitfaden für Sportsponsoring und Eventgestaltung, D 7.3.* Düsseldorf: Raabe Verlag.

Naul, R., & Großbröhmer, R. (1996). *40 Jahre Schulsport in Nordrhein-Westfalen: Lehrplantheorie und Unterrichtspraxis.* Düsseldorf: Concept.

Naul, R., & Neuhaus, W. (1996). Active lifestyle and physical fitness of German boys and girls aged 12 to 14. *International Journal of Physical Education, 33*(1), 27–36.

Naul, R., Neuhaus, W., & Rychtecky, A. (1996). Daily physical activity and motor performance of school youngsters, Czech/German study. In J. Chyracaková & M. Kohoutek (Eds.), *Sport kinetics 95* (pp. 392–399). Prague: FTVS.

Rychtecky, A., Naul, R., & Neuhaus, W. (1996). Physical activity and motor performance of Prague and Essen school youngsters. *Acta Universitatis Carolinae, Kinanthropologica, 32*(2), 5–21.

Naul, R. (1997). Comparative P.E. and sport studies in Germany in the 1990s. *International Journal of Physical Education, 34*, 80–102.

Naul, R. (Ed.). (1997). *Contemporary studies in the National Olympic Games movement.* Frankfurt am Main: Lang (Sport Science International, vol. 2).

Naul, R. (1997). The National Olympic Games Movement in Germany. In R. Naul (Ed.), *Contemporary studies in the National Olympic Games Movement* (pp. 95–116). Frankfurt am Main: Lang (Sport Sciences International vol. 2).

Naul, R. (1997). Sporting lifestyle, motor performance and Olympic ideals of youth in Europe. *ICSSPE/CIEPSS Bulletin*, 23, 40–41.

Naul, R. (1997). Zukünftige Ausstattungs- und Spielmöglichkeiten. In BISP (Ed.), *Planung und Bau von Hallen für Turnen, Spiele und Mehrzwecknutzung* (No. 1). Köln: BISP.

Naul, R., Pieron, M., Telama, R., & Almond, L. (1997). Etude du style de vie d´adolescents européens. Considérations théoriques, objectifs et méthodologie de recherche. *Sport*, 159/160, 43–50.

Naul, R., Pieron, M., Telama, R., Almond, L., & Rychtecky, A. (1997). *Sporting lifestyle, motor performance, and Olympic ideals of youth in Europe.* Essen: WGI.

Naul, R., Telama, R., & Rychtecky, A. (1997). Physical fitness and active lifestyle of Czech, Finnish and German youth. *Acta Universitatis Carolinae, Kinanthropologica, 33*(2), 5–15.

Naul, R. (1998). Anfänge des Fußballspiels an Rhein und Ruhr. In WFV (Ed.), *100 Jahre Fußball im Westen* (p. 12). Kassel: Agon.

Naul, R. (1998). Der Sportmediziner Ferdinand August Schmidt. In WFV (Ed.), *100 Jahre Fußball im Westen* (p. 18). Kassel: Agon.

Naul, R. (1998). Fußball als Stauchball, Kampfspiel und Rasenschach. In WFV (Ed.), *Programmheft 100 Jahre Fußball im Westen* (pp. 8–9). Duisburg: Westdeutscher Fußballverband.

Naul, R. (1998). German baseball. Ein "neues" Spiel - ein "altes" Spiel? *Lehrhilfen für den Sportunterricht*, 47, 43–48.

Naul, R., Hardman, K., Pieron, M., & Skirstad, B. (Eds.). (1998). *Physical activity and active lifestyle of children and youth.* Schorndorf: Hofmann (Sport Science Studies vol. 10).

Naul, R. (1998). Olympic ideals and Olympic education of youth. In R. Naul, K. Hardmann, M. Pieron, & B. Skirstad (Eds.), *Physical activity and active lifestyle of children and youth* (pp. 29–48). Schorndorf: Hofmann (vol. 10 Sport Science Studies).

Naul, R. (1998). Research report – Sporting lifestyle, motor performance, and Olympic ideal of youth in Europe. *ECSS News Bulletin*, 4, 20–21.

Naul, R. (1998). The impact of the Le Havre Congress on physical education in selected European countries. In N. Müller (Ed.), *Coubertin and Olympism questions for the future, coubertin and Olympism questions for the future* (pp. 141–153). Niedernhausen: Schors.

Naul, R. (1998). (Mitherausgeber). In WFV (Ed.), *100 Jahre Fußball im Westen.* Kassel: Agon.

Naul, R., Hardmann, K., & Telama, R. (1998). Sport pedagogy. In ICSSPE (Ed.), *Vade Mecum. Directory of Sport Science* (1st ed., pp. 33–38). Berlin: ICSSPE.

Naul, R., & Rychtecky, A. (1998). Vosprijatije olimpijskich idealov zanimajushchimisja i ne zanimajushchimisja sportom 12 i 15 letnimi detmi (The perception of Olympic ideals by children of 12–15 years of age who deal and who do not deal with sport). In V. Rodichenko (Ed.), *Proceedings of the international forum "youth-science- Olympism"* (pp. 204–207). Moscow: Sovietsky Sport.

Naul, R., Rychtecky, A., & Voulle, P. (1998). *Does the different state of physical fitness in young Europeans correspond with their moral assessment of Olympic Ideal?* (Paper presented at the 11th ISCPES Biannual Congress). Leuven.

Naul, R., & Telama, R. (1998). Sporting lifestyle, motor performance, and Olympic ideal of youth in Europe – Part two. *ICSSPE Bulletin*, 25, 40–41.

Naul, R. (1999). Dansk gymnastics europæiske dimension. In J. Hansen & N. Grinderslev (Eds.), *Idræt og samfund, krop og kultur. Idrætshistorisk Årbog* (pp. 35–49). Odense: Universitet.

Naul, R. (1999). Das Fach Sport im Berufskolleg - ein Beitrag zur umfassenden beruflichen Handlungskompetenz. In LSW (Ed.), *Sport im Berufskolleg, Werkstattberichte Curriculumrevision im Schulsport* (No. 8, pp. 14–30). Soest, LSW.

Naul, R. (1999). Future sports in the ballhouse. *Sport, Bäder, Freizeitbauten, 39*(4), 27–30.

Naul, R. (1999). Nationales Olympia und deutsche Kampfspiele. In M. Lämmer (Ed.), *Deutschland in der Olympischen Bewegung. Eine Zwischenbilanz* (pp. 25–35). Melsungen: Bernecker.

Naul, R. (1999). Willibald Gebhardt – "der deutsche Coubertin". In M. Lämmer (Ed.), *Deutschland in der Olympischen Bewegung. Eine Zwischenbilanz* (pp. 35–37). Melsungen: Bernecker.

Naul, R. (1999). Olympic education: An elective module within the European Master's degree programme in physical education: The European Network of Sport Science in Higher Education. *Acta Universitatis Carolinae, Kinanthropologica, 35*(1), 29–32.

Naul, R. (1999). Olympic education: What do Students know and think about Olympic Ideals. *Wychowanie fizyczne i sport*, 43(suppl. 1), 23–36.

Naul, R. (1999). School sport and competition: Sport pedagogy. In S. Bailey (Ed.), *The multidisciplinary series of physical education and sport series* (pp. 73–83). Oxford: Meyer & Meyer.

Naul, R. (1999). Sport pedagogy. *International Journal of Physical Education*, 36, 139–140.

Naul, R. (1999). Sporthalle: Sportnutzung ehemaliger Industriebauten – Beispiele aus der IBA Emscherpark, Deutschland. *Sportstättenbau und Bäderanlagen*, 33(5), 425–426.

Naul, R., & Lämmer, M. (Eds.). (1999). *Willibald Gebhardt: Pionier der Olympischen Bewegung (Schriftenreihe des Willibald-Gebhardt-Instituts; Band 3).* Aachen: Meyer & Meyer

Naul, R. (1999). Willibald Gebhardt – Leben und Werk des Pioniers der internationalen Olympischen Bewegung in Deutschland. In R. Naul & M. Lämmer (Ed.), *Willibald Gebhardt. Pionier der olympischen Bewegung* (pp. 11–29). Aachen: Meyer & Meyer (Schriftenreihe des Willibald-Gebhardt-Instituts, Band 3).

Naul, R. (1999). Willibald Gebhardt – Begründer der Olympischen Erziehung in Deutschland. In O. Grupe (Ed.), *Einblicke: Aspekte olympischer Sportentwicklung* (pp. 206–209). Schorndorf: Hofmann.

Telama, R., Nupponen H., Naul, R., Arvisto, M., Barabas, A., Rychtecky, A., & Pieron, M. (1999). Physical activity and lifestyle among young people. In P. Parisi, E. Pigozzi, & G. Prinzi (Eds.), *Proceedings of the 4th annual congress of the european college of sport science*, July 14–17, 1999 (p. 240). Rome: ECSS.

Telama, R., Nupponen, H., Naul, R., Pieron, M., & Rychtecky, A. (1999). *Goal orientation and physical activity in the context of lifestyle among adolescents; a cross-cultural comparison (Poster)*. Besancon: International AIESEP Congress.

Von 2000 bis 2009

Laakso, L., Nupponen, H., Telama, R., Arvisto, M., Naul, R., & Pieron, M., et al. (2000). Determinants of leisure time physical activity among youth in some European countries. In J. Avela, P. V. Komi, & J. Komuleinen (Eds.), *Proceedings of the 5th Annual Congress of the ECSS* (p. 66). Jyväskylä: Univ.

Naul, R. (2000). Das Fach Sport im Berufskolleg – Ein Beitrag zur umfassenden Handlungs-Kompetenz. In H. Aschebrock (Ed.), *Erziehender Schulsport. Pädagogische Grundlagen der Curricularrevision in Nordrhein-Westfalen* (pp. 205–220). Boenen: Kettler.

Naul, R. (2000). Quels criteres pour un Degré de Master en sciences du sport? In N. Midol & M. Astolfi (Eds.), *Les Sciences du sport entre cultures et techniques* (pp. 83–92). Gemenos: AFRAPS.

Naul, R. (2000). Sport- und Mehrzweckhallen der Zukunft. *Sport, Bäder, Freizeitbauten*, 40(5), 48–51.

Naul, R. (2000). Sportnutzung ehemaliger Industriebauten – Beispiele aus der IBA Emscher Park, Deutschland. In IAKS (Ed.), *16th International Congress Report 1999* (pp. 54–59). Köln: IAKS.

Naul, R. (2000). Sportnutzung ehemaliger Industriebauten. Beispiele aus der IBA Emscher Park, Deutschland. *Sportstättenbau und Bäderanlagen, 34*(1), 16–21.

Naul, R., Jonischeit, L., & Wick, U. (2000). *Turnen, Spiel und Sport in Schule und Verein: Jugendsport zwischen 1870 und 1932*. Aachen: Meyer & Meyer (Schriftenreihe des Willibald-Gebhardt- Instituts; Band 4).

Naul, R., & Okade, Y. (Eds.). (2000). *Sportwissenschaft in Deutschland und Japan*. Aachen. Meyer & Meyer (Schriftenreihe des Willibald-Gebhardt-Instituts; Band 5).

Naul, R., & Petry, K. (2000). Bachelor & Master – eine Einführung in das Schwerpunktthema. *dvs-informationen, 15*(3), 6.

Naul, R., & Petry, K. (2000). Flexibilisierung, Auslandsorientierung, Akkreditierung: Was kann die Sportwissenschaft aus der Studie zur Reform-Diskussion über Bachelor und Master-Studiengängen lernen? *dvs-informationen, 15*(3), 7–15.

Naul, R., & Rychtecky, A. (2000). Kenntnisse, Schulunterricht und Informationsquellen zum Thema Olympia bei deutschen und tschechischen Jugendlichen. In Deutsches Olympisches Institut (Ed.), *DOI Jahrbuch* (pp. 97–115). St. Augustin: Academia.

Naul, R. (2000). Olympische Ideale: Idee und Wirklichkeit aus der Sicht von europäischen Jugendlichen. In M. Messing, N. Müller, & H. Preuß (Eds.), *Blickpunkt Olympia. Entdeckungen, Erkenntnisse, Impulse* (pp. 175–190). Kassel: Agon.

Naul, R., & Rychtecky, A. (2000). The Olympic ideal of fair play: Moral dissonance between vision and practice? *ICSSPE Bulletin*, 30, 42–43.

Rychtecky, A., & Naul, R. (2000). Relexe Fair play u ceskych a nemeckych pubesentu a adolescentu. In Univerzita Karlova (Ed.), *Spolecenske Problemy Kinanthropologie, Spolecenske Problemy Kinanthropologie* (pp. 60–65). Praha: FTVS.

Naul, R. (2001). Concepts of physical education in Europe. In J. Mester (Ed.), *6th Annual Congress of the European College of Sport Science* (p. 193). Köln: Strauss.

Telama, R., Nupponen, H., Rychtecky, A., & Hoffmann, D. (2001). Physical and social determinants of perceived body competence in European children and adolescents. In J. Mester (Ed.), *6th Annual Congress of the European College of Sport Science* (p. 62). Köln: Strauss.

Rychtecky, A., & Naul, R. (2001). Lifestyle activities and participation in sports among and between Czech and German Adolescents. In J. Mester (Ed.), *6th Annual Congress of the European College of Sport Science* (p. 63). Köln: Strauss.

Naul, R. (2001). *Der Sportverein in Nordrhein-Westfalen: ein Versager, Bewahrer oder Förderer der motorischen und sportlichen Jugendarbeit?* Essen: WGI.

Naul, R. (2001). Der Sportverein in Nordrhein-Westfalen: ein Versager, Bewahrer oder Förderer der motorischen und sportlichen Jugendarbeit? In LSB NRW (Ed.), *"Jugendarbeit im Sportverein – Anspruch und WirklichkeitÐ. Expertenmaterial zur Brettschneider-Studie* (pp. 58–69). Duisburg: LSB.

Naul, R. (2001). Der Sportverein in NRW: Versager, Bewahrer oder Förderer der sportlichen Jugendarbeit? *Amtliche Mitteilungen des Westdeutschen Fußballverbandes*, 10, 15–18.

Naul, R. (2001). Der Sportverein in NRW: Versager, Bewahrer, oder Förderer der sportlichen Jugendarbeit? *Teil1. Amtliche Mitteilungen des Westdeutschen Fußballverbandes*, 9, 18–20, 23.

Naul, R. (2001). Fußballsport im Freizeit- und Leistungsprofil von Jugendlichen in Europa. In D. Teipel, R. Kemper, & D. Heinemann (Eds.), *Nachwuchsförderung im Fußball. 15. Jahrestagung der dvs-Kommission Fußball vom 17.-19.11.1999 in Jena* (pp. 147–164). Hamburg: Czwalina.

Naul, R. (2001). *Kenntnisse zum Thema Olympia und ihre sozialen Determinanten bei deutschen Jugendlichen. DOI-Forum "Olympische BewegungÐ.* (Abstrakt) Berlin: DOI.

Naul, R. (2001). Von der fiktiven IPES zur real existierenden ICSSPE-Studie. *sportunterricht*, 50, 144–147.

Naul, R. (2001). Wie fußball-aktiv sind Jugendliche und wie fit sind junge Fußballspieler? *Amtliche Mitteilungen des Westdeutschen Fußballverbandes*, 5, 15–16.

Naul, R. (2001). Willibald Gebhardt, Carl Diem und die Anfänge des olympischen Sports in Deutschland. In M. Krüger (Ed.), *Olympische Spiele: Bilanz und Perspektiven im 21. Jahrhundert* (pp. 74–89). Münster: Lit-Verlag.

Naul, R. (2001). Zu Unrecht vergessen. Öffentliche Anerkennung und Würdigung Willibald Gebhardts steht noch aus. *Sport in Berlin*, 51(9), 10.

Naul, R., Hoffmann, D., Rychtecky, A., Boschek, P., & Vuolle, P. (2001). *Measurement of Olympic values and virtues: A factor-model of Olympic ideals. Beijing* (Poster).

Naul, R., Nupponen, H., Telama, R., Rychtecky, A., Arvisto, M., & Pieron. (2001). *Physical activity as a predictor of motor fitness among European youth. Besancon* (Poster).

Rychtecky, A.& Naul, R. (2001). Hodnoceni olympijskych principu u ceske a nemecke mladeze. In P. Tillinger, A. Rychtecky, & T. Peric (Eds.), *Sport v Ceske republice na zacatku noveho tisicileti* (pp. 245–249). Praha: FTVS.

Naul, R., Rychtecky, A., Vuolle, P., & Hoffmann, D. (2001). Children's perception of Olympic champions: Bronze, silver and gold – I do not want? In ISAPA (Ed.), *13th international symposium, 5th European Congress adapted physical activity* (Abstract). Vienna.

Naul, R., Rychtecky, A., Vuolle, P., & Hoffmann, D. (2001). *Evaluation of teaching Olympic education: What do European adolescents know about the Olympic Movement. Pre-Congress Workshop 14th International Congress for Physical Education and Sport for Girls and Women* (Abstract). Athen.

Naul, R., & te Uhle, R. (Eds.). (2001). *Sportunterricht zwischen zentraler Steuerung und lokaler Autonomie.* (Schriftenreihe der EAdS, Band 16). Velen: eads.

Naul, R. (2002). Applied sport history: A new science for the sporting goods market. In K. E. A. Szikora (Ed.), *Proceedings of the 6th ISHPES Congress* (pp. 431–434). Budapest: Platin-Print Bt.

Naul, R. (2002). Concepts of physical education in Europe. *ICSSPE-Bulletin*, 34, 14–15.

Naul, R. (2002). Das Willibald-Gebhardt-Institut (WGI) – ein internationales Forschungsinstitut für Olympische Erziehung. In T. Jurka (Ed.), *Studia z Dziejow Kultury Fizycznej* (pp. 113–121). Gorzow: WLKP.

Naul, R. (2002). Der Sportverein in Nordrhein-Westfalen - Ein Versager, Bewahrer oder Förderer der motorischen und sportlichen Jugendarbeit (Auszüge). In LSB/SJ NRW (Ed.), *Jugendarbeit im Sportverein – Anspruch und Wirklichkeit. Informationen, Diskussionsbeiträge und Handlungsansätze zur Brettschneider-Studie* (pp. 26–29). Duisburg: LSB.

Naul, R. (2002). fußball ohne grenzen. *DFB-aktuell*, 8, 34–35.

Naul, R. (2002). Lebenslanges Lernen: Die Schlüsselrolle des Sports. In Akademie des Sports LSB Niedersachsen (Ed.), *Lebenslanges Lernen in Europa – Visionen und Chancen zum Europäischen Jahr der Bildung durch Sport im Jahr 2004* (pp. 17–23). Hannover: LSB.

Naul, R. (2002). Neuer Berufsschulsport im Berufskolleg in Nordrhein-Westfalen. In F. Brauweiler & P. Klingen (Ed.), *Berufsschulsport. Neue Wege gehen – Bewährtes integrieren* (pp. 13–25). Bielefeld: Bertelsmann Verlag.

Naul, R. (2002). Olympische Erziehung – Ein integriertes Konzept für Schule und Verein. In Düsseldorf Rhein-Ruhr 2012 GmbH (Ed.), *Weißbuch Olympische Erziehung* (pp. B1–B32). Düsseldorf: DRR GmbH.

Naul, R. (2002). Olympische Erziehung- Chancen und Aufgaben für den Schulsport. *sportunterricht*, 51, 267–272.

Naul, R. (2002). Rolle und Stellenwert der Sportwissenschaft in den Graduiertenkollegs und Innovationskollegs der Deutschen Forschungsgemeinschaft (DFG). *dvs-informationen*, 17(2), 4–6.

Loth, W., Naul, R., Gehrmann, S., & Schröer, C. (2002). Das DFG-Graduiertenkolleg "Europäische Gesellschaft" an der Universität Essen. *dvs-informationen*, 17(2), 12–17.

Naul, R. (2002). Über die Rationalisierung des Sportunterrichts. In W. Borgers (Ed.), *Tempel und Ringe zwischen Hochschule und Olympischer Bewegung, Tempel und Ringe zwischen Hochschule und Olympischer Bewegung* (pp. 109–119). Köln: Carl und Liselott Diem-Archiv.

Naul, R., & Hardman, K. (Eds.). (2002). *Sport and physical education in Germany.* London and New York: Routledge (Hardcover).

Naul, R., & Hardman, K. (Eds.). (2002). *Sport and physical education in Germany.* London and New York: Routledge (Paperback).

Naul, R. (2002). History of sport and physical education in Germany 1800–1945. In R. Naul & K. Hardman (Eds.), *Sport and physical education in Germany* (pp. 15-27). London and New York: Routledge.

Hardman, K., & Naul, R. (2002). Sport and physical education in the two Germanies. In R. Naul & K. Hardman (Eds.), *Sport and physical education in Germany* (pp. 28–76). London and New York: Routledge.

Naul, R., & Hardmn, K. (2002). Sport and physical education in re-unified Germany, 1990–2000. In R. Naul & K. Hardman (Eds.), *Sport and physical education in Germany* (pp. 77-86). London and New York: Routledge.

Naul, R. (2002). Physical education in schools. In R. Naul & K. Hardman (Eds.), *Sport and physical education in Germany* (pp. 87–98). London and New York: Routledge.

Naul, R. (2002). Physical education teacher training. In R. Naul & K. Hardman (Eds.), *Sport and physical education in Germany* (pp. 99–112). London and New York: Routledge.

Naul, R., & Lämmer, M. (Eds.). (2002). *Die Männer um Willibald Gebhardt: Anfänge der olympischen Bewegung in Europa.* Aachen: Meyer & Meyer (Schriftenreihe des Willibald-Gebhardt- Instituts; Band 7).

Rychtecky, A., & Naul, R. (2002). Hodnoceni olympijskych principu: komparativni studie ceske a nemecke mladeze. In A. Suchomel & R. Antos (Eds.), *Telesna vychova a sport 2002* (pp. 239–243). Liberec Euregion Nisa: Technika Univerzita.

Rychtecky, A., & Naul, R. (2002). Lifestyle activities and participation in sports among Czech and German Youth. *Acta Universitatis Carolinae – Kinanthropologica, 32*(2), 5–21.

Telama, R., Naul, R., Nupponen, H., Rychtecky, A., & Vuolle, P. (2002). *Physical fitness, sporting lifestyles and Olympic ideals: Cross-cultural studies on youth sport in Europe*. Schorndorf: Hofmann (Sport Science Studies vol. 11).

Naul, R. (2003). Besprechung: Riordan & Jones (Eds.), Sport and physical education. *European Physical Education Review*, 9, 319–321.

Naul, R. (2003). Concepts of physical education in Europe. In K. Hardman (Ed.), *Physical education: Deconstruction and reconstruction; issues and directions* (pp. 35–52). Schorndorf: Hofmann (Sport Science Studies, vol. 12).

Naul, R. (2003). Der Olympische Gedanke auf dem Prüfstand? Bedeutung der Olympischen Erziehung für den Sportverein. In *Europäisches Jahr der Erziehung durch Sport 2004* (pp. 23–42). Hannover: LSB.

Naul, R. (2003). Der Sport der Heranwachsenden im internationalen Vergleich. In W. Schmidt (Ed.), *Erster Deutscher Kinder- und Jugendsportbericht*, (pp. 361–379). Schorndorf: Hofmann.

Naul, R. (2003). Die integrierte Olympische Erziehung: Ein Konzept für Schule und Sportverein. *Stadion*, 29, 287–300.

Naul, R. (2003). Koncepe skolni telesne vychovy v Evrope. *Ceska Kinanthropologie*, 7, 39–53.

Naul, R. (2003). Offene Ganztagsschule – Chancen und Risiken für den Sport. In W. &. W. LSB NRW (Ed.), *Bewegung, Spiel, Sport und Ganztagsbetreuung in Schulen* (pp. 9–25). Duisburg: WFLV.

Naul, R. (2003). Olympische Erziehung – "olympisch leben". *Alpheios*, 4, 36–39.

Naul, R. (2003). PISA und der Schulsport. *sportunterricht*, 52, 131.

Naul, R., Hoffmann, D., Telama, R., & Nupponen, H. (2003). PISA-Schock auch im Schulsport? Wie fit sind deutsche und finnische Jugendliche? *sportunterricht*, 54, 137–141.

Naul, R. (2003). Stichwort: "Willibald Gebhardt". In P. Röthig & R. Prohl (Eds.), *Sportwissenschaftliches Lexikon* (7th ed., pp. 213–214). Schorndorf: Hofmann.

Naul, R. (2003). Was halten deutsche Jugendliche vom Fair Play als Olympisches Ideal? In E. Bertke, H. Kuhn, K. Lennartz, & M. Lämmer (Eds.), *Olympisch bewegt. Festschrift zum 60. Geburtstag von Prof. Dr. Manfred Lämmer* (pp. 247–252). Köln: DSHS.

Naul, R., & Gustmann, H. (2003). Note Zehn: Olympische Erziehung als Alleinstellungsmerkmal der Düsseldorfer Bewerbung. *Alpheios*, 4, 29–32.

Naul, R., & Hoffmann, D. (2003). *Grensoverschrijdende sportcontacten in de Euregio.* Velen: eads (Schriftenreihe der EAdS, No. 22).

Naul, R., & Hoffmann, D. (2003). *Grenzüberschreitender Sportaustausch in der Euregio* Velen: eads (Schriftenreihe der EAdS, No. 21).

Naul, R., Okade, Y., & te Uhle, R. (Eds.). (2003). *4. Deutsch-Japanisches Symposium zur Sportwissenschaft.* Velen: eads (Schriftenreihe der EAdS, No. 20).

Binder, D., & Naul, R. (2004). Transnational applications of Olympism in school programs: A comparison of North American and European educational orientations. In *Proceedings of the 2004 Pre-Olympic Congress* (p. 208). Thessaloniki: Aristotle University.

Naul, R. (2004). Children's perceptions of Olympic champions: A "good example" as a Personal Role model? In: International Olympic Academy (Ed.), *Report of the I.O.A.'s Special Sessions and Seminars 2003* (pp. 624–628). Athen: Eptalofos S.A.

Naul, R. (2004). Der olympische Gedanke auf dem Prüfstand: Allgemeine Sportentwicklung und Konsequenzen für die Erziehung junger Sportler. In Sportakademie des LSB Thüringen (Ed.), *Die Bedeutung der olympischen Ideale für die Erziehung* (pp. 6–15). Erfurt: LSB.

Naul, R. (2004). Die Dekade für Olympischen Erziehung an Rhein und Ruhr. In H. Gustmann & M. Lämmer (Eds.), *Olympic Spirit. Reader zur der Vortragsreihe "Olympic Spirit" der Düsseldorf Rhein-Ruhr 2012 GmbH* (pp. 72–87). Köln: GFP.

Naul, R. (2004). Fair play as perceived by youngsters. In G. Gandolfi (Ed.), *Sport, Ethics, cultures* (pp. 249–259). Rapallo: Penthalon.

Naul, R. (2004). I giovani e il Fair Play. In G. Gandolfi (Ed.), *Sport. Etiche. Culture. vol. III* (pp. 249–259). Rapallo: Penthalon.

Naul, R. (2004). Fair Play jako olimpijski ideal i jako wzor osobistego zachowania w sporcie. Protrzeba a rzeczywistosc - opinie mlodych Europejczykow. In Z. Zukowksa & R. Zukowksi (Eds.), *FAIR PLAY w europejskiej kulturze i edukacji* (pp. 118–129). Warszawa: Estrella.

Naul, R. (2004). "Olympia" – ...und was wissen deutsche Jugendliche darüber? J. Buschmann, K. Lennartz & St. Wassong (Eds.), *Spiel – Spiele – Olympische Spiele* (pp. 101–111). Aachen: Meyer & Meyer (Schriftenreihe des Willibald-Gebhardt-Instituts, Band 10).

Tazuke, S., Okade, Y., & Naul, R. (Eds.). (2004). *Globalisierung des Sports: Zur Rolle der japanischen und deutschen Sportwissenschaft.* Kyoto: Tadas Book Ltd.

Naul, R. (2004). Olympic education in Germany: Reviewed from a global perspective. In S. Tazuke, Y. Okade, & R. Naul (Eds.), *Globalisierung des Sports: Zur Rolle der japanischen und deutschen Sportwissenschaft* (pp. 23–31). Kyoto: Tadas Book Ltd.

Naul, R. (2004). *Olympische Erziehung – alter Hut oder neue Aufgabe für den Schulsport? Betrifft: Sport, 26*(4), 15–20.

Naul, R. (2004). Olympische Erziehung: Alter Hut oder neue Aufgabe für den Schulsport? (gekürzte Fassung). In B. Rapp & G. Krapp (Eds.), *Der Olympische Gedanke – und die Welt des Sports. Lehrerheft* (pp. 26–28). Rot an der Rot: Krapp & Gutknecht.

Naul, R. (2004). Von der Pädagogik des Olympismus zur Didaktik der Olympischen Erziehung. In R. Geßmann (Ed.), *Olympische Erziehung – eine Herausforderung an Sportpädagogik und Schulsport* (pp. 113–130). Sankt Augustin: Academia.

Naul, R. (2004). Do today's children need traditional games or future sports? In G. Pfister (Ed.), *Games of the past – Sports for the future* (pp. 157–162). Sankt Augustin: Academia (ISHPES-Studies, vol. 9).

Naul, R. (Ed.). (2004). *Nachwuchsförderung im Kinder- und Jugendfußball in Europa: Ergebnisbericht der 17. Jahrestagung der dvs-Kommission Fußball vom 21–23. November 2001 in Duisburg.* Aachen: Meyer & Meyer (Schriftenreihe des Willibald-Gebhardt-Instituts; Band 9).

Naul, R. (2004). Sind unsere Kinder und Jugendlichen körperlich "zu dick" und moralisch "zu dünn"? In Deutscher Sport Bund (Ed.), *DSB Jahresmagazin 2004/2005* (pp. 76–77). Frankfurt/M.: DSB.

Naul, R., Brüggemann, D., Völz, C., Halemeier, M., Altenbeck, B., & Hoffmann, D. (2004). Talente motorisch vielseitig fordern und fördern: Teil 2: Fußball-Fitness – wie gut sind die Grundlagen für das Techniktraining? *fußball training, 11*(8), 18–21.

Naul, R., & Krotee, M. (2004). German legacy: From German gymnastics to physical education in modern unified Germany. In A. R. Hofmann (Ed.), *Turnen and sport. Transatlantic transfers* (pp. 69–84). Münster: Waxmann (German & American Studies in Sport, vol. 5).

Naul, R., Völz, C., & Brüggemann, D. (2004). Talente motorisch vielseitig fordern und mental fördern: Teil 1: Ohne erweiterte Fitness keine technischen Spitzenleistungen. *fußball training, 22*(7), 34–39.

Naul, R. Brüggemann, D., Völz, Ch., Halemeier, M., Altenbeck, B., & Hofmann, D. (2004). Talente motorisch vielseitig fordern und mental fördern: Teil 2: Fußball-Fitness – wie gut sind die Grundlagen für das Techniktraining? *fußball training, 22*(8), 18–21.

Naul, R., Völz, C., Hoffmann, D., Altenbeck, B., Brüggemann, D., & Halemeier, M. (2004). Talente motorisch vielseitig fordern und mental fördern: Teil 3: Sind junge Fußballtalente auch psycho-mental auf der Höhe? *fußball training, 22*(10), 33–41.

Naul, R., Völz, C., & te Uhle, R. (Eds.). (2004). Deutsch-Niederländisches Schulsportsymposium. Velen: eads (Schriftenreihe der EAdS, No. 18).

Naul, R. (2004). Die Entwicklung des Schulsports in Nordhein-Westfalen – Fachdidaktische Ansätze und Lehrpläne. In R. Naul, C. Völz & R. te Uhle (Eds.), *Deutsch-Niederländisches Schulsportsymposium* (pp. 1–21). Velen: eads.

Brettschneider, W.-D., & Naul, R. (2005). Omtrent studie levensstijl van jongeren – Sedentariteit: Analyse van de rol van sport in de context van onderwijs en als middel om het evenwicht te herstellen. *Tijdschrift voor Lichamelijke Opvoeding, 208*(4), 12–18.

Brettschneider, W.-D., & Naul, R. (2005). Physical activity and fitness in relation to cardiovascular diseases in young people. *European Health and Fitness Association*, 1, 9–10.

Brettschneider, W.-D., & Naul, R. (2005). Study on young people's lifestyles and sedentariness and the role of sport in the context of education and as a means of restoring the balance. Retrieved from http://ec.europa.eu/sport/documents/lotpaderborn.pdf

Naul, R. (2005). Bewegung, spiel und sport im offenen Ganztag. *sportunterricht*, 54, 68–72.

Naul, R. (2005). Der Deutsche Fußball-Fitness-Test. In BDFL (Ed.), *Internationaler Trainer-Kongress 18. bis 20. Juli 2005. Düsseldorf "Aktuelle Trainingsansätze zurÐ. Umsetzung einer modernen Spielphilosophie* (pp. 62–70). Frankfurt/M.: BDFL.

Naul, R. (2005). Fair play as an Olympic ideal and a personal behaviour pattern in sports activities: Vision and reality assessed by young European students. PLAY FAIR! *Academic Supplement, 3*(1), 1–3.

Naul, R. (2005). Olympische Erziehung in Europa. In Würth, S., et al. (Eds.), *Sport in Europa. 17. Sportwissenschaftlicher Hochschultag der Deutschen Vereinigung für Sportwissenschaft* (p. 74). Hamburg: Czwalina.

Richter, C., & Naul, R. (2005). Psycho-mentale Eigenschaften von jungen Fußball-Spielern: Das Essener Fußball-Persönlichkeits-Inventar. In: Würth, S., et al. (Eds.), *Sport in Europa. 17. Sportwissenschaftlicher Hochschultag der Deutschen Vereinigung für Sportwissenschaft* (p. 269). Hamburg: Czwalina.

Richter, C., te Uhle, R., & Naul, R. (2005). (Eds.) *Qualitätsentwicklung im Schulsport. 3. Europäisches Schulsportforum.* Velen: eads (Schriftenreihe der EAdS, No. 24).

Naul, R. (2005). Qualitätsstandards für den us-amerikanischen Sportunterricht. In C. Richter, R. te Uhle, & R. Naul (Eds.), *Qualitätsentwicklung im Schulsport. 3. Europäisches Schulsportforum* (pp. 173–186). Velen: EAdS.

Naul, R. (2005). Talente im Fußballsport: motorisch vielseitig fordern und mental fördern: Fußballfitness und psycho-mentale Persönlichkeitsmerkmale bei D-Junioren im Kader des Westfälischen Fußball- und Leichtathletik-Verbandes (FLVW). In J. Mester & S. Knuth (Eds.), *Spitzensport im Jahr der Olympischen Sommerspiele 2004* (pp. 77–118). Aachen: Meyer & Meyer.

Naul, R. (2005). Zur Olympischen Erziehung in Deutschland. In S. Cys, P. Olesniewicz, W. Starosciak, & D. Kurz (Eds.), *Wychowanie fizyczne w polkich i niemieckich szkolach, wspolczesne trendy - Schulsport in polnischen und deutschen Schulen* (pp. 97–107). Wroclav: AWF.

Naul, R., Boetes, B., Geßmann, R., Richter, C., & Grabs, R. (2005). *Olympische Werteerziehung in Schule und Sportverein: Handreichung mit Materialien für Multiplikatoren, Moderatoren und Bildungsreferenten für den Schul- und Vereinssport.* Essen: WGI.

Naul, R., & Brettschneider, W.-D. (2005). Young people's lifestyles and sedentariness in Europe. *Acta Universitatis Carolinae, Kinanthropologica, 41*(2), 25–34.

Naul, R., & Völz, C. (2005). Bewegung, Spiel und Sport im Ganztag: pädagogische Ansprüche zwischen Schulsport und Kinder- und Jugendhilfe. In A. Gogoll & A. Menze-Sonneck (Eds.), *Qualität im Schulsport* (pp. 73–78). Hamburg: Czwalina.

Naul, R., & Wick, U. (2005). Schulprojekte (2004-2006). In Fußball- und Leichtatletikverband Westfalen & Willibald Gebhardt Institut (Eds.), *Fußballregion Ruhrgebiet* (pp. 247–250). Göttingen: Die Werkstatt.

Naul, R., & Wick, U. (2005). Stichwort football. In R. Bartlett, C. Gratton, & C. G. Rolf (Eds.), *Encyclopedia of International Sport Studies* (pp. 540–543). London and New York: Routledge.

Richter, C., & Naul, R. (2005). Psycho-Mental qualities of young soccer players. *Acta Universitatis Carolinae, Kinanthropologica, 41*(2), 119–125.

Rychtecky, A., & Naul, R. (2005). Goal orientation and perception of Olympic ideals in Czech and German Youth. *Acta Universitatis Carolinae, Kinanthropologica, 41*, 35–48.

Naul, R. (2006). Der Auf- und Abstieg des Fußballspiels an Schulen im Spiegel der Lahrpläne von Nordrhein-Westfalen. In M. Raab, A. Arnold, K. Gärtner, J. Köppen, C. Lempertz, & N. Tielemann, et al. (Eds.), *Zukunft der Sportspiele: fördern, fordern, forschen* (pp. 61–64). Flensburg: University Press.

Naul, R. (2006). Der Fußball als "Schule des Lebens : über "männliche und "weibliche Spielweisen und das gute alte "teutsche Kampfspiel. *männerforum*, 34, 6–9.

Naul, R. (2006). *Drei Jahre Bewegung, Spiel und Sport im Ganztag – Aufgaben und Perspektiven. Zukunftsdialog II der Fachverbände im LandesSportBund NRW* (pp. 18–26). Duisburg: LSB.

Naul, R. (2006). Olympische Erziehung in Schule und Sportverein in Europa. *Leipziger Sportwissenschaftliche Beiträge, 47*(2), 30–31.

Naul, R. (2006). Olympische Werte als soziale Basiskompetenz für gutes Regieren. In W. Tokarski (Ed.), *Sportpolitik. Theorie- und Praxisfelder von Governance im Sport* (pp. 121–134). Köln: Strauß.

Naul, R. (2006). *Pierre de Coubertin und Eugenio Brunetta d`Usseaux: Kongress der Nationalen Olympischen Akademie von Italien in Pinerolo. Alpheios, (7*, 79–80.

Naul, R. (2006). Qualität auch im Kinder- und Jugendsport! In Sportjugend NRW (Ed.), *Die Zukunft des Kinder- und Jugendsports* (pp. 77–80). Duisburg: Rhiem.

Naul, R. (2006). *Sozialpädagogische und sportpädagogische Professionen im Ganztag: Qualitätskriterien und Fortbildungsbausteine für Angebote mit Bewegung, Spiel und Sport: Expertise für das BLK Verbundprojekt "Lernen für den GanzTagĐ.* Essen: Universität.

Naul, R. (2006). Visions of football – International conference for the 2006 FIFA World Cup TM, Munich, ICC, 27–29 July, 2005 (Bericht). *Sportwissenschaft, 26*(1), 94–99.

Brettschneider, W.-D., Naul, R., Bünnemann, A., & Hoffmann, D. (2006). Übergewicht und Adipositas bei Kindern und Jugendlichen: Ernährungsverhalten, Medienkonsum und körperliche (In-)Aktivität im europäischen Vergleich. *Spectrum der Sportwissenschaften, 18*(2), 25–45.

Naul, R., & Hoffmann, D. (2006). Healthy children in sound communities: An Euregional community setting project. In P. Heikinaro-Johansson, R. Telama, & E. McEvoy (Eds.), *The role of physical education and sport in promoting physical activity and health. AIESEP World Congress, 5–8 July 2006* (p. 106). Jyväskylä, Finland: Kopijyvä Oy.

Naul, R., & Richter, C. (2006). All-Day schools: New challenges for physical education and extra-curricular physical activities. In P. Heikinaro-Johansson, R. Telama, & E. McEvoy (Eds.), The role of physical education and sport in promoting physical activity and health. *AIESEP World Congress, 5–8 July 2006* (p. 214). Jyväskylä, Finnland: Kopiyvä Oy.

Naul, R., & Richter, C. (2006). Olympic education in the Netherlands and in Germany a cross-border curricular project. In P. Heikinaro-Johansson, R. Telama, & E. McEvoy (Eds.), *The role of physical education and sport in promoting physical activity and health. AIESEP World Congress, 5–8 July 2006* (p. 196). Jyväskylä, Finnland: Kopiyvä Oy.

Naul, R., & Richter, C. (2006). Integrierte Olympische Erziehung – ein Konzept für Schule und Verein. In W. Buss, S. Güldenpfennig, & A. Krüger (Eds.), *Zur Neubegründung der Olympischen Idee. Denkanstöße* (pp. 83–110). Wiesbaden: Stumm.

Naul, R., Richter, C., Harlemeier, M., Bertelsmeier, F., & Brüggemann, D. (2006). Praxis und Demonstration trainingsbegleitender Maßnahmen im Fußball. Maßnahmen zur Behebung physischer Defizite. In K. Weber (Ed.), *Wissenschaftlicher Transfer für die Praxis der Sportspiele. Gemeinsames Symposium der dvs-Kommissionen Sportspiele, Fußball und Tennis vom 18–20. November 2004 an der Deutschen Sporthochschule Köln* (pp. 143–151). Köln: Strauß.

Naul, R., Richter, C., Hoffmann, D., Altenbeck, B., Brüggemann, D., Fischer, B., et al. (2006). Fußball-Fitness und Psycho-Mentale Persönlichkeitsmerkmale bei DJunioren im Kader des Fußball- und Leichtathletik-Verbandes Westfalen (FLVW). In K. Weber (Ed.), *Wissenschaftlicher Transfer für die Praxis der Sportspiele. Gemeinsames Symposium der dvs-Kommissionen Sportspiele, Fußball und Tennis vom 18–20. November 2004 an der Deutschen Sporthochschule Köln* (pp. 132–143). Köln: Strauß.

Wick, U., & Naul, R. (2006). *Fußballregion Ruhrgebiet, Ausstellung & Event 2005/2006. Ausstellungsdokumentation.* Essen: WGI.

Brettschneider, W.-D., & Naul, R. (Eds.). (2007). Obesity in Europe: Young people's physical activity and sedentary lifestyles (Sport Science International, vol. 4). Frankfurt am Main: Lang.

Brettschneider, W.-D., & Naul, R. (2007). Obesity in Europe. Young peoples's physical activity and sedentary lifestyles. In W.-D. Brettschneider & R. Naul (Eds.), *Obesity in Europe. Young people's physical activity and sedentary lifestyles* (pp. 7–26). Frankfurt am Main: Lang.

Brettschneider, W.-D., Naul, R., Bünnemann, A., & Hoffmann, D. (2007). Obesity, physical activity patterns and sedentary lifestyles of young people in Austria, Germany and Switzerland. In W.-D. Brettschneider & R. Naul (Eds.), *Obesity in Europe. Young people's physical activity and sedentary lifestyles* (pp. 281–323). Frankfurt am Main: Lang.

Eulering, J., Naul, R., Schilling, K. A., & Wick, U. (2007). Bewegung, Spiel, Rhythmus, Tanz: Sportkultur 2010. In K. A. Schilling (Ed.), *Kulturmetropole Ruhr. Perspektivplan II* (pp. 447–486). Duisburg: WAZ.

Krüger, M., & Naul, R. (2007). 9. Internationale Session für Direktoren der Nationalen Olympischen Akademien. *Sportwissenschaft*, 37, 346–349.

Naul, R. (2007). Der PISA-Schock und seine Einflüsse auf die Entwicklung der Sportpädagogik in Deutschland. In D. Teipel, R. Kemper, & Y. Okade (Eds.), *Topics of social and behavioral science in sport. 6th German-Japanese Symposium 19–21. 09. 2005 in Jena* (pp. 21–36). Köln: Strauß.

Naul, R. (2007). Fair play in football: International concepts of projects and campaigns. In M. Lämmer & A. Grassi (Eds.), *12th European Fair Play Congress* (pp. 37–49). Cologne/ Udine: EFPM.

Naul, R. (2007). Historical development of physical education in Europe. In J. Tolleneer, M. Lenoir, & W. Laporte (Eds.), *Physical education: Past, present, and future. Symposium "100 years of physical education and movement sciencesÐ at Ghent Universiteit* (pp. 13–27). Ghent: Universiteit.

Naul, R. (2007). Olympische Erziehung: Kernaufgabe Nationaler Olympischer Akademien. *Newsletter der Österreichischen Olympischen Akademie*, 1, 2–3.

Naul, R. (2007). *Olympische Erziehung (Edition Schulsport, Band 7).* Aachen: Meyer & Meyer.

Naul, R. (2007). Socialpaedagogiske udfordringer, sport og idraet i born og unges institutionaliserede hverdagsliv. *Tidsskrift for Socialpaedagogik*, 20, 11–22.

Naul, R. (2007). Talentförderung im Mädchenfußball: Wege und Probleme. In G. dawietz & U. Kraus (Eds.), *Die Zukunft des Fußballs ist weiblich. Beiträge zum Frauen- und Mädchenfußball* (pp. 90–107). Aachen: Meyer & Meyer.

Naul, R., & Hoffmann, D. (2007). Healthy children in sound communities: A Euregional community setting project. In P. Heikinaro-Johansson, R. Telama, & E. McEvoy (Eds.), *The role of physical education and sport in promoting physical activity and health. AIESEP World Congress, 5–8 July 2006, Jyväskylä, Finland* (pp. 258–267). Jyväskylä: Kopiojyvä Oy.

Naul, R., & Richter, C. (2007). All-Day schools: New challenges for physical education and extra-curricular physical activities. In P. Heikinaro-Johansson, R. Telama, & E. McEvoy (Eds.), *The role of physical education and sport in promoting physical activity and health. AIESEP World Congress, 5–8 July 2006, Jyväskylä, Finland* (pp. 104–109). Jyväskylä: Kopiyvä Oy.

Naul, R., & Hoffmann, D. (2007). *Projekthandbuch "Gesunde Kinder in gesunden Kommen" Planungsgrundlagen für ein euro-kommunales Netzwerk zur Förderung eines aktiven Lebensstils von Kindern und Jugendlichen.* Essen: Willibald-Gebhardt-Institut.

Boetes, B., Naul, R., & Richter, C. (2008). Olympische educatie op school en bij vereinigingen. *Lichamelijke Opvoeding, 96*(3), 8–10.

Naul, R. (2008). Bejing Olympic games bid books 2000 and 2008: What role does Olympic Education Play? In ISCPES (Ed.), *Proceedings of the 16th Conference of the International Society for Comparative Physical Education and Sport. Global merging. A new era for sport and physical education. 23–26 July 2008* (Abstract, p. 13). Macau: ISCPES.

Naul, R. (2008). In Sportschuhen durch den Ganztag. In Innenministerium NRW & Hessisches Ministerium des Innern und für den Sport (Ed.) *Kinder l(i)eben Sport.* (pp. 74–81). Düsseldorf: Innenminsterium.

Naul, R. (2008). Die Ganztagsschule: neues Lernen in der Schule für Kopf und Körper, mit Bewegung und Verstand. In W. Schmidt, R. Zimmer & K. Völker (Eds.), *Zweiter Deutscher Kinder- und Jugendsportbericht* (pp. 319–335). Schorndorf: Hofmann.

Naul, R. (2008). *Olympic education.* Oxford: Meyer & Meyer.

Naul, R. (2008). Physical education – Sport education – Olympic education: From historical deconstruction to future reconstruction? In FIEP (Ed.), *Physical education in change. FIEP World Congress, August 6–10, 2008* (Abstract, p. 72). Vierumäki: FIEP.

Naul, R. (2008). The Bejing Olympic games bid books 2000 and 2008: What role does Olympic education play? In ICSSPE (Ed.), Education and medicine in sport. Sport Sciences and human society in the 21th century. *International Convention on Science*, August 1–5, 2008 (Abstract, p. 218). Guangzhou: ICEMIS.

Naul, R. (2008). The five steps of the IOC for the promotion of Olympic education. In Renmin University of China (Ed.), *Great breakthrough. Olympic theories and practices in the east* (pp. 294–296). Beijing: Renmin University.

Naul, R., Brüggemann, D., & Völz, C. (2008). Talented players need physical and mental challenges: Part 1: Optimizing performance with total fitness. *Success in Soccer*, 11(4), 28–33.

Naul, R., Brüggemann, D., & Völz, C. (2008). Talented players need physical and mental challenges: Part 2: Soccer fitness: Laying the foundation for technique training. *Success in Soccer, 11*(5), 16–19.

Naul, R., Geßmann, R., & Wick, U. (2008). *Olympische Erziehung in Schule und Verein: Grundlagen und Materialien*. Schorndorf: Hofmann.

Naul, R. (2008). Olympic pedagogy as a theory of development of ethical and humanistic values in education. *Sporto Mokslas, 3*(53), 9–15.

Naul, R., Hoffmann, D., Altenbeck, B., Brüggemann, D., Völz, C., & Harlemeier, M. (2008). Talented players need physical and mental challenges: Part 3: Psychology: Assessing youth players' mental skills. *Success in Soccer, 11*(6), 30–38.

SpOGATA (2008). *Abschlussbericht der Essener Pilotstudie im Rahmen des Evaluationsprojektes "Evaluation des BeSS-Angebotes an offenen Ganztgasgrundschulen im Primarbereich in seinen Auswirkungen auf die Angebote und Struktur von Sportvereinen, Koordinierungsstellen und die Ganztagsförderung des Landessportbundes NRW in Nordrhein-Westfalen.* Essen: WGI.

Naul, R. (2009). Der Doppelauftrag der Olympischen Erziehung im Schulsport. *Alpheois, 9*, 62–67.

Naul, R. (2009). Bewegung, Spiel und Sport in offenen Ganztagsgrundschulen: Ausgewählte Evaluationsergebnisse der Essener Pilotstudie. In DSJ/DOSB (Ed.), *Fachkonferenz Sport & Schule 2009* (pp. 36–37). Frankfurt/M: DOSB.

Naul, R., & Wick, U. (Eds.) (2009). *20 Jahre dvs-Kommission Fußball. Herausforderung fürden Fußballsport in Schule und Sportverein.* Hamburg: Czwalina (Beiträge und Analysen zum Fußballsport XVI).

Naul, R. (2009). 20 Jahre dvs-Kommission Fußball. In R. Naul & U. Wick (Eds.), *20 Jahre dvs-Kommission Fußball. Herausforderung für den Fußballsport in Schule und Sportverein* (pp. 11–16). Hamburg: Czwalina.

Naul, R. (2009). Disputa di calcio: Die Heidelberger Ballschule – diskutiert aus Sicht eines zeitgemäßen Fußballsports – Einführung. In R. Naul & U. Wick (Eds.), *20 Jahre dvs-Kommission Fußball* (pp. 205–206). Hamburg: Czwalina.

Naul, R., Krüger, A., & Schmidt, W. (Eds.). (2009). Kulturen des Jugendsports. Bildung, Erziehung und Gesundheit. Aachen: Meyer & Meyer (Schriftenreihe des Willibald-Gebhardt-Instituts, Band 13).

Brettschneider, W. D., Hoffmann, D., Naul, R., & Steinzen, A. (2009). Von den Erkenntnissen aus dem EU-Bericht "Young people's lifestyle and sedentariness" zur handlungsorientierten Umsetzung der Empfehlungen im Rahmen von zwei lokalen Präventionsstudien. In R. Naul, A. Krüger, & W. Schmidt (Eds.), *Kulturen des Jugendsports. Bildung, Erziehung und Gesundheit* (pp. 67–104). Aachen: Meyer & Meyer.

Geis, S., Hoffmann, D., Naul, R., & Wick, U. (2009). Bewegung, Spiel und Sport an offenen Ganztagsgrundschulen in Essen. In: R. Naul, A. Krüger & W. Schmidt (Eds.), *Kulturen des Jugendsports. Bildung, Erziehung und Gesundheit* (pp. 149–166). Aachen: Meyer & Meyer.

Hoffmann, D., & Naul, R. (2009). Die körperliche und motorische Entwicklung von Grundschulkindern in der Gemeinde Velen im Rahmen der Pilotstudie des Interventionsprojekts "Gesunde Kinder in gesunden Kommunen" (gkgk). In R. Naul, A. Krüger, & W. Schmidt (Eds.), *Kulturen des Jugendsports. Bildung, Erziehung und Gesundheit* (pp. 105–128). Aachen: Meyer & Meyer.

Naul, R., & Geßmann, R. (2009). Olympische Erziehung - seit wann, wozu und für. Wen. In: R. Naul, A. Krüger, & W. Schmidt (Eds.), *Kulturen des Jugendsports. Bildung, Erziehung und Gesundheit* (pp. 39–64). Aachen: Meyer & Meyer.

Naul, R. (2009). Das Willibald-Gebhardt-Institut: Von der Idee zur Institutsgründung – 20 Jahre Forschung und Entwicklung (1989–2009). In R. Naul, A. Krüger, & W. Schmidt (Eds.). *Kulturen des Jugendsports. Bildung, Erziehung und Gesundheit* (pp. 283–298). Aachen: Meyer & Meyer.

Naul, R. (2009). Die Entwicklung der historischen Systeme für die schulische Leibeserziehung in Europa. In R. Naul, A. Krüger, & W. Schmidt (Eds.), *Kulturen des Jugendsports. Bildung, Erziehung und Gesundheit* (pp. 131–148). Aachen: Meyer & Meyer.

Naul, R., & Lämmer, M. (Eds.). (2009). *Willibald Gebhardt. Pionier der Olympischen Bewegung.* Aachen: Meyer & Meyer (2. Auflage).

Naul, R. (2009). Social-pedagogical challenges and sport in the institutionalised daily lives of children and young people. In J. Kornbeck & N. R. Jensen (Eds.), *The diversity of social pedagogy in Europe* (pp. 120–143). Bremen: Europäischer Hochschulverlag.

Naul, R. (2009). Address on behalf of the lectures of the session. In International Olympic Academy (Ed.), *Proceedings of the 9th international session for directors of proceedings of the 9th international session for Directors of National Olympic Academies* (pp. 294–296). Athens: Liviani.

Naul, R. (2009). Olympic pedagogy as a theory of development of ethical and humanistic values in education. In International Olympic Academy (Ed.), *Proceedings of the 9th international session for directors of national Olympic academies* (pp. 27–40). Athens: Livani.

Naul, R. (2009). Zusammenfassung AK 3 "Talentsuche in der Schule. In L. Nordmann, C. Dolche, & H. Steingrebe (Red.), *Sport ist Spitze. Talente suchen und entwickeln -Chancen, Wege und Methoden* (pp. 127–128). Aachen: Meyer & Meyer.

Naul, R., & Schmidt, W. (2009). Der Stellenwert der Sportspiele in Schule und Hochschule In: H. F. Voigt & G. Jendrusch (Eds.), *Sportspielforschung in Bochum. Was war, was ist und was sein könnte* (pp. 31–41). Hamburg: Czwalina.

Richter, C., & Naul, R. (2009). Versagensangst und Selbstvertrauen als psychomentale Merkmale in Relation zur Fußball-Fitness bei D- und C-Junioren des Fußball und Leichtathletikverbandes Westfalen (FLVW). In U. Frick, (Ed.), *Fußball in Schule und Verein. Eine Herausforderung für Forschung und Lehre* (pp. 99–106). Hamburg: Czwalina.

Naul, R., & Richter, C. (2009). Die Entwicklung und Profilbildung der Fußball-Fitness bei D-/ C-Junioren des Verbandskaders des Fußball- und Leichtathletikverbandes Westfalen

(FLVW). In: Frick, U. (Ed.), *Fußball in Schule und Verein. Eine Herausforderung für Forschung und Lehre* (pp. 90–98). Hamburg: Czwalina.

Von 2010 bis 2018

Naul, R., Tietjens, M., Geis, S., & Wick, U. (2010). Bewegung, Spiel und Sport im Ganztag von NRW - Konzept und Ergebnisse der Essener Pilotstudie. In P. Böcker & R. Laging (Eds.), *Bewegung, Spiel und Sport in der Ganztagsschule: Schulentwicklung, Sozialraumorientierung und Kooperationen* (pp. 143–158). Hohengehren: Schneider.

Naul, R. (2010). Buchbesprechung von R. Geßmann "Richtlinien und Lehrpläne für den Schulsport. Eine kommentierte Dokumentation 1945–2008" *sportunterricht, 59*, 20–22.

Naul, R., Behnke, & Schmelt, D. (2010). Local networks for active living in healthy communities: A Dutch-German cross-border approach and its diffusion into four other European countries. *Gymnica. Acta Universitatis Palackianae Olomucensis, 40*(3), 56.

Naul, R. (2010). Der Fußballsport in Schule und Hochschule – zwischen Bildung und Ausbildung. In O. Höner, R. Schreiner, & F. Schultz (Ed.), *Aus- und Fortbildungskonzepte im Fußball* (pp. 15-31). Hamburg: Czwalina.

Naul, R. (2010). Conceptual diversity and future directions of physical education in the global context. In Japanese Society of Sport Education (Ed.), *Proceedings of the international conference for the 30th anniversary of the Japanese Society of Sport Education – "Quality of Sport in Schools and out of Schools"* (pp. 1–13). Tokyo: Ministry of Education, Culture, Sport, Science and Technology.

Naul, R, Schmelt, D., & Hoffmann, D. (2010). Gesunde Kinder in gesunden Kommunen (gkgk) - ein deutsch-niederländisches Interventionsprojekt zur Förderung eines aktiven Lebensstils bei Kindern im Grundschulalter. In Chuo Univ. (Ed.), *Proceedings 7. deutsch-japanisches Symposium zur Sportwissenschaft* (pp. 27–29). Tokyo: Chuo Univ.

Naul, R., Kornbeck, J., & L'Hoir, M. (2010). Multi-actor strategies of health-enhanced physical activities for children in local communities of the European Union. *Medicine & Science in Sports & Exercise, 42*(5), S253–S254.

Naul, R. (2010). *Olympic education*. Oxford: Meyer & Meyer Sport Ltd. (2nd corrected edition).

Naul, R. (2010). Didactic approaches to teaching Olympic education. In V. Girginov (Eds.), *The Olympics. A critical reader* (pp. 405–415). London and New York: Routledge.

Naul, R. (2010). Rezension von R. Geßmanns "Richtlinien und Lehrpläne für den Schulsport". Eine kommentierte Dokumentation 1945–2007. Köln: Strauß 2008. *sportunterricht, 59*, 20–22.

Naul, R. (2011). Gastkommentar: Olympische Erziehung: eine Aufgabe für die Jugendarbeit im Fußballverein? *WestfalenSport 6*, 18.

Dreiskämper, D., Schmelt, D., Hoffmann, D., & Naul, R. (2011). "Gesunde Kinder in gesunden Kommunen" (gkgk) – ein kommunales Interventionsprojekt zwischen Schule, Sportverein und Kommune zur Förderung eines aktiven Lebensstils. In *Kreativität – Innovation – Leistung. Wissenschaft bewegt SPORT bewegt Sportwissenschaft* (p. 113). 20. Sportwissenschaftlichen Hochschultag der dvs, Halle/Saale. Hamburg: Czwalina.

Naul, R., Wick, U., Geis, S., & Gerhardt, A. (2011). Fokus: Sportvereine – Evaluation von Bewegungs-, Spiel- und Sportangeboten an Offenen Ganztagsschulen in. Nordrhein-Westfalen. In E. Adelt, D. Fiegenbaum, S. & Rinke (Eds.), *Bewegen im Ganztag. Bewegung, Spiel und Sport im Ganztag* (pp. 12–17). Münster: ISA.e.V.

Naul, R. (2011). Value education in schools form a cross-curricular perspective. In J. Dovalil, et al. (Eds.), *Fair play education in schools: A shared responsibility* (pp. 18–29). Prague: Czech Olympic Committee.

Naul, R. (Ed.) (2011). *Bewegung, Spiel und Sport in Ganztagsschulen. Bilanz und Perspektiven.* Aachen: Meyer & Meyer (Edition Schulsport, Band 14).

Naul, R. (2011). Ganztägiges Lernen mit Turnen, Spiel und Sport – historische Entwicklungslinien zwischenSchule und Sportverein. In R. Naul (Ed.), *Bewegung, Spiel und Sport in Ganztagschulen. Bilanz und Perspektiven* (pp. 30–50). Aachen: Meyer & Meyer.

Naul, R. (2011). Der Bildungsauftrag von Bewegung, Spiel und Sport in Ganztagsschulen. In R. Naul (Ed.), *Bewegung, Spiel und Sport in Ganztagschulen. Bilanz und Perspektiven* (pp. 76–97). Aachen: Meyer & Meyer.

Schmelt, D., Hoffmann, D., & Naul, R. (2011). Bewegungsbildung und Gesundheitsförderung in der Ganztagsschule. In R. Naul (Ed.), *Bewegung, Spiel und Sport in Ganztagschulen. Bilanz und Perspektiven* (pp. 115–132). Aachen: Meyer & Meyer.

Wick, U., Naul, R., Geis, S., & Tietjens, M. (2011). Essener Sportvereine und der offene Ganztag. In R. Naul (Ed.), *Bewegung, Spiel und Sport in Ganztagschulen. Bilanz und Perspektiven* (pp. 229–250). Aachen: Meyer & Meyer.

Naul, R.(2011). *HCSC – Final report. Technical implementation.* Essen: WGI/dsj.

L'Hoir, M., Kloeze, E., Boere-Bonnekamp, M., De Jong, S., Nawijn, L., & Naul, R. (2011). GKGK, *een community based project ter bevordering van gezond gedrag. Tijschrift Jeugdgezondheitszorg, 43*(4), 82–83.

Naul, R. (2011). Conceptual diversity and future directions of physical education in the global context. *Japanese Journal of Sport Education Studies, 30*(2), 39–50.

Naul, R. (2011). Community-based local partnerships to revitalize PE and health: The German case. In East China Normal University & West Virgina University (Eds.), *Physical education and sport: Challenges and future directions* (pp. 50–51). Shanghai: East China Normal University.

Naul, R., Schmelt, D, & Hoffmann, D. (2011). Bewegungsförderung in der Schule. In Landesinstitut für Gesundheit und Arbeit des Landes NRW (Ed.), *Gesundheit durch Bewegung. Empfehlungen für Wissenschaft und Praxis* (pp. 70–73). Düsseldorf: Liga.

Naul, R. (2011). Internationalism in Olympic education? A European comparison. In H. Preuss & K. Liese (Eds.), *Internationalism in the Olympic Movement. Idea and reality between nations, cultures, and people* (pp. 63–74). Wiesbaden: VS-Verlag.

Naul, R. (2011). Value education in schools from a cross-curricular perspective. *Sporto Mokslas, 3*(56), 49–56.

Kloeze, E, Boere-Boonekamp, M. M., de Jong, S., Nawijn, L., Bakker, I., Naul, R., & L'Hoir, M. P. (2011). Slaap, voeding en beweging in relatie tot overgewicht. *Voeding NU, 13*(3), 10–13.

Naul, R., & Holze. J. (2011). Sports development and young people: The role of international organizations. In B. Houlihan & M. Green (Eds.), *Routledge handbook of sports development* (pp. 198–211). London and New York: Routledge.

L`Hoir, M. P., Vlasblom, E., Boere-Boonekamp, M. M., van Dommelen, P., Schmelt, D., Dreiskämper, D., & Naul, R. (2012). Slaap en overgewicht in Europees onderzoek. Gezonde Kinderen in een Gezonde Kindomgeving (GKGK). *Tijdschrift voor Jeugdgezondheidszorg, 44*(6), 102–104.

Naul, R., & Wick, U. (2012). Die pädagogische Dimension der Olympischen Idee. *Alpheios 10*, 83–87.

Höfer, A., Kluge, V., Lennartz, K., & Naul, R. (2012). Gebhardts Vermächtnis oder wie viel Gebhardt steckt in Diem? In U. Wick & A. Höfer (Eds.), *Willibald Gebhardt und seine Nachfolger. Sport und Olympia in Deutschland bis 1933* (pp. 19–32). Aachen: Meyer & Meyer.

Edginton, C. R., Chin, M. K., & Naul, R. (2012). The global forum for health and physical education pedagogy: A new social movement. *International Journal of Physical Education, XLIX*(3), 31–39.

Naul, R., Edginton, C., & Chin, M. K. (2012). Global forum for physical education 2012. *Global Journal of Health and Physical Education Pedagogy, 1*(2), 166–168.

Naul, R., Schmelt, D., Dreiskämper, D., & Hoffmann, D. (2012). Gesunde Kinder in gesunden Kommunen (gkgk) – ein deutsch-niederländisches Interventionsprojekt zur Förderung eines aktiven Lebensstils bei Kindern im Grundschulalter. *WGI newsletter, 4*, 3.

Naul, R., Schmelt, D, Dreiskämper, D., & Hoffmann, D. (2012). Healthy children in sound communities (gkgk) – A German-Dutch intervention project to promote an active lifestyle for children. *WGI-Newsletter, 4*, 2.

Naul, R., Schmelt, D., Hoffmann, D., & Dreiskämper, D. (2012). "Gesunde Kinder in gesunden Kommunen" (gkgk). Ein deutsch-niederländisches Interventionsprojekt zur Förderung eines aktiven Lebensstils. In T. Ichiba (Ed.), *Das 7. Deutsch-Japanische Sportwissenschaftliche Symposium* (pp. 135–158). Tokyo: Hiltop Press.

Forschungsgruppe SpOGATA. (2012). *Abschlussbericht des Forschungsprojektes "Evaluation des BeSS-Angebotes an offenen Ganztagsschulen im Primarbereich in seinen Auswirkungen auf die Angebote und Struktur von Sportvereinen, Koordinierungsstellen und die Ganztagsförderung des Landessportbundes NRW in Nordrhein-Westfalen"*. Essen/Münster: WGI.

Naul, R. (2012). Olympic pedagogy – A new discipline of research and teaching at Olympic Study Centres? In Univ. of Loughborough (Ed.), *The first inaugural colloquium of Olympic study and research centres* (pp. 11–12). Loughborough: University.

Naul, R. (2012). Value education at schools. In R. Meesuen, J. Duchateau, B. Roelands, M. Klass, B. de Geus, S. Baudry, & E. Tsolakidis (Eds.), *17th annual Congress of the ECSS. Book of abstracts* (p. 459). Ghent: Bloso.

Naul, R., & Schmelt, D. (2012). Gesunde Kinder in gesunden Kommunen (gkgk) – Ein europäisches Interventionsprojekt zur Förderung eines aktiven Lebensstils bei Kindern in Grundschulalter. In I. Hunger & R. Zimmer (Ed.), *Frühe Kindheit in Bewegung. Entwicklungspotenziale nutzen* (pp. 192–205). Schorndorf: Hofmann.

Wick, U., Naul, R., Tietjens, M., & Gerhardt, A. (2012). Evaluation of physical activities. Games and sports at open all-day primary schools in North Rhine Westphalia. *WGI-Newsletter, 4,* 1.

Naul, R., Schmelt, D., Dreiskämper, D., & Hoffmann, D. (2012). Healthy children in sound communities (gkgk) – A German-Dutch intervention project to promote an active lifestyle for children. *WGI-Newsletter, 4,* 2.

Naul, R. (2012). Global Forum 2012. Revitalizing health and physical education through commonly-based networks. *WGI-Newsletter, 3,* 1–2.

Naul, R., Dreiskämper, D., & Schmelt, D. (2012). *Healthy children in sound communities. Abschlussbericht über die deutschen Kommunen im EU Interventionsprojekt 2010-2011 (Preparatory Actions).* Essen: WGI.

Naul, R., Schmelt, D., & Hoffmann, D. (2012). Bewegungsförderung in der Schule – was wirkt? In G. Geuter & A. Hollederer (Ed.), *Handbuch Bewegungsförderung und Gesundheit* (pp. 229–246). Bern: Huber.

Naul, R., Schmelt, D., Dreiskaemper, D., Hoffmann, D., & L'Hoir, M. (2012). "Healthy children in sound communities" (HCSC/gkgk) – A Dutch-German community-based network project to counteract obesity and physical inactivity. *Family Practice, 29*(1), 110–116.

Naul, R. (2012). European Union multisector strategies to enhance health, physical education, and physical activities for children and youth. *Global Journal of Health and Physical Education Pedagogy, 1(*1), 22–41.

Edginton, C., Ming-Kai, Ch., Naul, R., & Herring, M. C. (2013). Global forum for physical education pedagogy 2012: Recommendations for community-based networking. *Asian Journal of Exercise and Sports Sciences, 10*(1), 1–15.

Graf, C., Beneke, R., Bloch, W., Bucksch, J., Dordel, S., Eiser, S., & Naul, R. (2013). Vorschläge zur Förderung der körperlichen Aktivität von Kindern und Jugendlichen in Deutschland. *Monatsschrift Kinderheilkude,161*(5), 439–446. doi:10.1007/s00112-012-2863-

Dreiskaemper, D., Tietjens, M., Freund, A., Honermann, S., & Naul, R. (2013). Validierung eines Fragebogens zum psychischen Selbstkonzept im Kindesalter PSK. In O. Stoll, A. Lau & S. Moczall (Eds.), *Angewandte Sportpsychologie. 45. Jahrestagung* asp (p. 120). Hamburg: Feldhaus.

Naul, R., Wick, U., & Gerhardt, A. (2013). Sport in der ganztägigen Bildung. In H. Aschebrock & G. Stibbe (Eds.), *Didaktische Konzepte für den Schulsport* (pp. 306–340). Aachen: Meyer & Meyer.

Naul, R. (2013). Netzwerke – wie funktionieren sie effektiv? (abstract) *Deutsche Zeitschrift für Sportmedizin, 64*(5), 143.

Naul, R., & Dreiskämper, D. (2013). Gesunde Kinder in gesunden Kommunen (gkgk). Kommunal vernetzte Gesundheitsförderung zwischen Schule und Sportverein. *Schule NRW, 65*(7), 324–326.

Naul, R., & Schmelt, D. (2013). Innovation oder Muß? Kommunale Gesundheitsförderung. *Brennpunkte der Sportwissenschaft, 34*, 33–41.

Naul, R. (2013). 20 years of international research in children and youth sport. *WGI-Report 2010–2012*, 6–12.

L'hoir, M., Tetteroo, S., & Boere-Boonekamp, U. A. (2013). Physical activity, nutrition, screen time and sleep associated with body weight and physical condition in young children. A pilot study of a cross-border project. *Sportwissenschaft, 43(*2), 116–123.

Naul, R. (2013). Bewegung, Spiel und Sport im Ganztag – Bewegung ist Bildung! *Betrifft Sport, 34*(H.4), 10–16.

Dreiskämper, D., Naul, R., & l'Hoir, M. (2013) Healthy children in healthy communities – Concept, strategy and results. In *TAFISA World Congress*. Enschede, Netherlands: NISB n.p.

Naul, R. (2013). Der preußische Leitfaden für den Turnunterricht von 1862. *Brennpunkte der Sportwissenschaft, 35*, 67–78.

Dreiskaemper, D., & Naul, R. (2013). *Was hängt mit motorischen Leistungen und BMI im Projekt "Gesunde Kinder in gesunden Kommunen (gkgk)Ð zusammen? Sportwissenschaft grenzenlos?!* (p. 240). Abstractband zum 21. dvs Hochschultag. Hamburg: Czwalina.

Utesch, T., Zentgraf, K., Tietjens, M., Strauss, B., & Naul, R. (2013). Eine Online-Datenbank zur webbasierten Erfassung von Entwicklungsdaten im Rahmen des Projekts GKGK. In Mess

Filip, Gruber Markus & Woll Alexander (Eds.), *21.* Sportwissenschaftlicher Hochschultag – Sportwissenschaft grenzenlos?! (p. 241). Hamburg: Czwalina.

Utesch, T., Naul, R., Dreiskämper, D., Strauss, B., & Tietjens, M. (2013). Die Bildung eines Motorik-Index mit Hilfe von Rasch Modellen. In Mess Filip, Gruber Markus & Woll Alexander (Eds.), *21. Sportwissenschaftlicher Hochschultag – Sportwissenschaft grenzenlos?!* (p. 242). Hamburg: Czwalina.

Naul, R., & Wick, U. (2014). Gkgk Arbeitshilfe. Essen: WGI.

Dreiskaemper, D., Naul, R., & Tietjens, M. (2013). Healthy children in sound communities' – Motor development intervention program in primary schools. In *The ISSP 13th World Congress of Sport Psychology*. Beijing, China. (n.p.).

Dreiskämper, D., & Naul, R. (2014). Gesunde Kinder in gesunden Kommunen. *Betrifft Sport, 36*(6), 10–16.

Graf, C., Beneke, R., Bloc, N. R., et al. (2014). Recommendations for promoting physical activity for children and adolescents in Germany. A consensus statement. *Obesity Facts, 7,* 178–190.

Naul, R. (2014). The European Union and Olympism: "Coubertin plan" and youth olympics. *Journal of Olympic History, 22*(2), 53–59.

Shen, J., Naul, R., You u. a. (2014). Research on Sino-European Health Project "healthy children in sound community". *Sport Science Research (China), 35*(6), 90–94.

Utesch, T., Strauss, B., & Tietjens, U. A. (2014). Using Rasch measurement to model a physical ability scale for German six- to nine-year-old students. *Journal of Sport & Exercise Psychology, 36*(supplement), 75. NASPSPA 2014 Conference, Minneapolis, USA.

Dreiskämper, D., & Naul, R. (2014). 'Healthy children in sound communities' - Motor development in intervention program in primary schools. In AIESEP World Congress, Auckland, Neuseeland (n.p.).

Naul, R., Dreiskämper, D., & Hoffmann, D. (2014). Physical and health education in Germany: From school sports to local networks for healthy children in sound communities. In M.-K. Chin & C. Edginton (Eds.), *Physical education and health – Global perspectives and best practice* (pp. 191–204). Urbana, IL: Sagamore.

Naul, R. (2014). Bewegung, Spiel und Sport in Ganztagsschulen: Konzepte für die Bildungspartnerschaft zwischen Ganztagsschulen und Sportvereinen. In H. Aschebrock, E. Beckers, R. P. Pack (Eds.), *Bildung braucht Bewegung* (pp. 221–249). Aachen: Meyer & Meyer.

Dreiskämper, D., & Naul, R. (2014) Motorische Entwicklung, Selbstkonzept und Gruppenklima im Interventionsprojekt, Gesunde Kinder in gesunden Kommunen'. In R. Frank, et al. (Eds.), *Performing under pressure* (p. 161). Hamburg: Czwalina.

Naul, R., Dreiskämper, D., & Hoffmann, D. (2014). Healthy children in sound communities (HCSC) – A German-Dutch intervention project to develop an active lifestyle for children in primary schools. *WGI-Newsletter, 5*, 1–4.

Naul, R., Dreiskämper, D., & Hoffmann, D. (2014). Gesunde Kinder in gesunden Kommunen (gkgk)ein deutsch-niederländisches Interventions-Projekt zur Förderung eines aktiven Lebensstils bei Kindern im Grundschulalter. *WGI-Newsletter, 6*, 1–4.

Naul, R., Dreiskämper, D., & Hoffmann, D. (2014). Zdrowe dzieci w zdrowych gminach (ZDZG) – niemiecko-holenderski projekt interwencyjny mający na celu wsparcie aktywnego stylu życia dzieci w wieku szkolnym. *WGI-Newsletter, 7*, 1–4.

Naul, R., Dreiskämper, D., & L'Hoir, M. (2014) Gezonde Kinderen in een Gezonde Kindomgeving (gkgk): Een Duits-Nederlands samenwerkingsproject gericht op het ontwikkelen van een actieve leefstijl bij kinderen in de leeftijd van 6 tot 10 jaar. *WGI-Newsletter, 8*, 1–4.

Vlasboom, E., L'Hoir, M., van Dommelen, P., Dreiskämper, D., Naul, R. (2014). Healthy children in sound communities in the Netherlands. In Univ. of Zurich (Ed.), *10th annual meeting and 5th conference of HEPA Europe* (p. 86). Zurich: Institute of Social and Preventive Medicine.

Naul, R., Heemsoth, S., & Dreiskämper, D. (2014) Healthy children in sound communities – The Influence of a Multi-Component Intervention on Overweight and Obese Children. In Univ. of Zurich (Ed.), *10th annual meeting and 5th conference of HEPA Europe* (pp. 86–87). Zurich: Institute of Social and Preventive Medicine.

Dreiskämper, D., l'Hoir, M., & Naul, R. (2014). Healthy children in sound communities' – A longitudinal intervention and control study about BMI and physical fitness development of primary school children in the Dutch municipality of Cuijk. In Univ. of Zurich (Ed.), *10th annual meeting and 5th conference of HEPA Europe* (pp. 87–88). Zurich: Institute of Social and Preventive Medicine.

Shen, J., Dreiskämper, D., Naul, R., & You, S. (2014). Healthy children in sound communities (HCSC CN-DE): A German and Chinese obesity intervention programme for primary school children. In Univ. of Zurich (Ed.), *10th annual meeting and 5th conference of HEPA Europe* (p. 88). Zurich: Institute of Social and Preventive Medicine.

Dreiskämper, D., Tietjens, M., & Honemann, U. A. (2015). PSK-Kinder – Ein Fragebogen zur Erfassung des physischen Selbstkonzepts von Kindern im Grundschulalter. *Zeitschrift für Sportpsychologie, 2015(*3), 97–111.

Naul, R. (2015). Pedagogical concepts for Olympic educatioin and didactic approaches to teaching Olympic education. In V. Girginov (Ed.), *Olympic studies, vol. III* (pp. 297–329). London and New York: Routledge.

Rehaag, R., Ferrari, N., Fischer, S. Fröhlich-Gildhoff, K., Graf, Ch., Naul, R., et al. (2015). Vernetzte kommunale Gesundheitsförderung für Kinder – wirkungsorientierte Qualitätssicherung und Verbesserung der Zugangschancen vulnerabler Zielgruppen. *Gesundheitswesen 77 – A152.* doi:10.1055/s-0035-1563108

Naul, R. (2015). Mehr dicke Kinder, weniger Bewegungszeiten, mehr Sitzstunden - und an allem ist die Ganztagsschule schuld? Der 3. Deutsche Kinder- und Jugendsportbericht zwischen Dichtung und Wahrheit. *Sportunterricht, 64*, 329–332.

Utesch, T., Strauss, B., Büsch, D., Zinner, J., Naul, R., Dreiskämper, D., Seidel, I., & Tietjens, T. (2015). Investigating the differentiation and stability of the concept of motor abilities across childhood. In *FEPSAC 2015*. Bern: Schweiz. (n.p.).

Naul, R., & Cloes, M (2015). School- and community-based PE and HEPA networks for daily physical activities. *Revista Espanola de Educacion Fisica y Deportes, LXVII*(No. 410 suppl.), 464.

Dreiskämper, D., & Naul, R. (2015). Healthy children in sound communities' – Four cohorts of longitudinal intervention project in primary schools. *Revista Espanola de Educacion Fisica y Deportes, LXVII(*No. 410 supl.), 466.

Naul, R., & Neuber, N. (2015). Zentrale Ergebnisse und Empfehlungen. In Forschungsgruppe SpOGATA (Ed.), *Evaluation der Bewegungs-, Spiel- und Sportangebote an Ganztagsschulen in Nordrhein-Westfalen* (pp. 215–220). Aachen: Meyer & Meyer.

Heemsoth, S., Tietjens, M., Dreiskämper, D., & Naul, R. (2015). Welche Rahmenbedingungen fördern und behindern eine gesundheitlich-motorische Entwicklung von Grundschulkindern im Ganztag? In Forschungsgruppe SPOGATA (Ed.), *Evaluation der Bewegungs-, Spiel- und Sportangebote an Ganztagsschulen in Nordrhein-Westfalen* (pp. 201–214). Aachen: Meyer & Meyer.

Heemsoth, S., Tietjens, M., & Naul, U. A. (2015). Bewegungsbildung und Gesundheitsförderung in offenen Ganztagsgrundschulen. In Forschungsgruppe SpOGATA (Ed.), *Evaluation der Bewegungs-, Spiel- und Sportangebote an Ganztagsschulen in Nordrhein-Westfalen* (pp. 137–163). Aachen: Meyer & Meyer.

Naul, R., Wick, U., & Geis, S. (2015). Die Bewegungs-, Spiel- und Sportangebote an Ganztagsgrundschulen. In Forschungsgruppe SpOGATA (Ed.), *Evaluation der Bewegungs-, Spiel- und Sportangebote an Ganztagsschulen in Nordrhein-Westfalen* (pp. 23–46). Aachen: Meyer & Meyer.

Naul, R., Neuber, N., & Tietjens, M. (2015). *Einleitung. In Forschungsgruppe SpoGATA Evaluation der Bewegungs-, Spiel- und Sportangebote an Ganztagsschulen in Nordrhein-Westfalen* (pp. 11–20). Aachen: Meyer & Meyer.

Naul, R., & Wick, U. (2015). Das Engagement der Sportvereine im offenen Ganztag. In Forschungsgruppe SpOGATA (Ed.), *Evaluation der Bewegungs-, Spiel- und Sportangebote an Ganztagsschulen in Nordrhein-Westfalen* (pp. 47–73). Aachen: Meyer & Meyer.

Naul, R. (2015). Die Ergebnisse der StEG-Studie und ihre Vergleichbarkeit mit den Ergebnissen der SpOGATA-Studie. In Forschungsgruppe SpOGATA (Ed.), *Evaluation der Bewegungs-, Spiel- und Sportangebote an Ganztagsschulen in Nordrhein-Westfalen* (pp. 105–113). Aachen: Meyer & Meyer.

Gerhardt, A., Wick, U., & Naul, R. (2015). Die individuellen BeSS-Anbieter an Ganztagsgrundschulen. In Forschungsgruppe SpOGATA (Ed.), *Evaluation der Bewegungs-, Spiel- und Sportangebote an Ganztagsschulen in Nordrhein-Westfalen* (pp. 75–89). Aachen: Meyer & Meyer.

Naul, R. (2015). Wie wird der Sport im Ganztag zu einem Erfolgsmodell? In Forschungsgruppe SpOGATA (Ed.), *Evaluation der Bewegungs-, Spiel- und Sportangebote an Ganztagsschulen in Nordrhein-Westfalen* (pp. 221–223). Aachen: Meyer & Meyer.

Wick, U., & Naul, R. (2015). Die Träger und Koordinierungsstellen des offenen Ganztags. In Forschungsgruppe SpOGATA (Ed.), *Evaluation der Bewegungs-, Spiel- und Sportangebote an Ganztagsschulen in Nordrhein-Westfalen* (pp. 91–103). Aachen: Meyer & Meyer.

Utesch, T., Dreiskämper, D., & Naul, R. (2015). Die gkgk-Online-Datenbank zum Monitoring der gesundheitlich-motorischen Entwicklung von Kindern und Jugendlichen. *WGI-Newsletter, 9,* 1-4.

Naul, R., Dreiskämper, D., & L'Hoir, M. (2015). Niederlande und Großbritannien – Befunde und Intervention. In W. Schmidt, N. Neuber, T. Rauschenbach, H. P. Brandel-Bredenbeck, J. Süssenbach, & C. Breuer (Ed.), *Dritter Deutscher Kinder- und Jugendsportbericht* (pp. 503–524). Schorndorf: Hofmann.

Naul, R., Schöckel, Y., & Dreiskämper, D. (2016). *Qualitätsentwicklung ZI:EL – Bericht zur Gesamtevaluation und wissenschfatlichen Begleitung des Projekts "Zukunftsinvestition: Entwicklung jungen Engagements im SportÐ.* Essen/Münster: WGI.

Naul,R., Schöckel, Y., & Dreiskämper, D. (2016). *Qualitätsentwicklung ZI:EL – Bericht zur Gesamtevaluation und wissenschfatlichen Begleitung des Propjekts "Zukunftsinvestition: Entwicklung jungen Engagements im SportÐ. Executive Summary.* Essen/Münster: WGI.

Shen, J., Naul, R., Dreiskämper, D, & Heemsoth, S. (2015). Investigation resesarch on health promotion and physcial activity intervention of primary school students – Take two Shanghai Yangpu District Schools as Examples. *Sport (Chinese), 120*(16), 54–55, and 118.

Utesch, T., Dreiskämper, D., Geukes, K., & Naul, R. (2016). Effekte des Zusammenspiels von tatsächlicher und wahrgenommener motorischer Leistung auf physische Aktivität: Eine Response Surface Analyse. In B. Halberschmidt, D. Dreiskämper, et al. (Eds.). *Spitzenleistungen und Sportpsychologie: Der Weg zu Olympia. 48. Jahrestagung der Arbeitsgemeinschaft für Sportpsychologie vom 05. bis 07. Mai, 2016 in Münster* (p. 98). Münster: WWU Münster.

Naul, R. (2016). Die Spiele und ihre Idee. Aufgaben und Ziele der Olympischen Erziehung. In Deutsche Olympische Akademie (Ed.),*Olympia ruft: mach mit! Basiswissen Olympische Spiele* (pp. 54–58). Frankfurt: DOA.

Utesch, T., Dreiskämper, D., & Naul, R. (2016). Monitoring motor development: Measuring physical fitness via an online data-base. In *European Network for the Promotion of Health-enhancing Physical Activity, HEPA 2016*. Belfast, Nordirland (n.p.).

Utesch, D., Dreiskämper, D., Strauss, B., & Naul, R. (2016). Longitudinal measurement of physical fitness in primary school: The developmental validity of identical item sets. In *SCAPPS 2016*. Waterloo, Kanada. (n.p.).

Naul, R., Schöckel, Y., & Dreiskämper, D. (2016). *Qualitätsentwicklung ZI:EL – Abschlussbericht zur Gesamtevaluation und wissenschaftlichen Begleitung des Projekts "Zukunftsinvestition: Entwicklung jungen Engagements im Sport der Deutschen Sportjugend."* Münster: WGI.

Naul, R. (2016). *Orinkku Kyoiku*. Tokyo: Taishukaan (Übersetzung von Olympic education ins Japanische).

Utesch, T., Dreiskämper, D., Geukes, K., & Naul, R. (2016). Effekte des Zusammenspiels von Fitness, motorischer Kompetenz und sportlicher Selbstwahrnehmung auf physische Aktivität: Eine Response Surface Analyse. *DGPS*, (n.p.).

Dreiskämper, D., Utesch, T., & Naul, R. (2016). Healthy children in sound communities' – Four cohorts of longitudinal intervention project in primary schools. *In NASPSPA 2016 Conference*. Montreal, Kanada (n.p.).

Utesch, T., Dreiskämper, D., Strauss, B., & Naul, R. (2017). The development of the physical fitness construct across childhood. *Scandinavian Journal of Medicine & Science in Sport, 27.* doi:10.1111/sms.12889

Dreiskämper, D., Utesch, T., & Naul, R. (2017). Motorische Diagnostik und gesundes Aufwachsen im Kindesalter. In *Jahrestagung der Gesellschaft für Public Health "Armut und Gesundheit"*. Berlin (n.p.).

Dreiskämper, D., Utesch, T., & Naul, R. (2017) Motorische Diagnostik und gesundes Aufwachsen im Kindesalter. In *Innovation & Technologie im Sport* (p. 89). Abstractband zum 23. Sportwissenschaftlichen Hochschultag der dvs, München. Hamburg: Czwalina.

Utesch, T., Dreiskämper, D., Naul, R., & Geukes, K. (2017). Eine realistische Selbsteinschätzung der motorischen Leistungsfähigkeit fördert die gesunde Entwicklung insbesondere von Kindern aus Risikogewichtsgruppen. In *49. Jahrestagung der Arbeitsgemeinschaft für Sportpsychologie* (asp). Bern, CH (n.p.).

Naul, R., Binder, D., Rychtecky, A., & Culpan, I. (Eds.). (2017). *Olympic education. An international review.* London and New York: Routledge (Hardcover).

Naul, R., Binder, D., Rychtecky, A., & Culpan, I. (2017). Introduction. In R. Naul, D. Binder, A. Rychtecky, & I. Culpan (Eds.), *Olympic education. An international review* (pp. 1–5). London and New York: Routledge.

Naul, R., Binder, D., Rychtecky, A., & Culpan, I. (2017). Olympic education and the Olympic movement: The role of Olympic Games, the International Olympic Committee, National Olympic Committees and National Olympic Academies in Olympic education. In R. Naul, et al. (Eds.), *Olympic education. An international review.* (pp. 351–356). London and New York: Routledge.

Naul, R., Binder, D., Culpan, I., & Rychtecky, A. (2017). Olympic education as 'I' education sportive': Olympic education as a part of physical education, extra-curricular physical activities, sport development and Olympic athlete training. In R. Naul, et al. (Eds.), *Olympic education. An international review.* (pp. 338–343). London and New York: Routledge.

Naul, R., Binder, D., Rychtecky, A., & Culpan, I. (2017). Olympic education as an academic study: Research, teching and training in higher learning institutes. In R. Naul, et al. (Eds.), *Olympic education. An international review.* (pp. 344–350). London and New York: Routledge.

Binder, D., & Naul, R. (2017). Olympic education as pedagogy: Terminology, pedagogical orientations and Olympic values education. In R. Naul, et al. (Eds.), *Olympic education. An international review* (pp. 331–337). London and New York: Routledge.

Psimopoulos, C., Binder, D., Vermillion, A., & Naul, R. (2017). United States of America: The Olympic Games as promoter of Olympic education: Issues and institution. In R. Naul et al. (Eds.), *Olympic education. An international review* (pp. 291–304). London and New York: Routledge.

Naul, R., Krüger, M., Geßmann, R., & Wick, U. (2017). Germany: Formal Olympic education at schools and informal Olympic learning in sport clubs. In R. Naul, et al. (Eds.), *Olympic education. An international review* (pp. 177–191). London and New York: Routledge.

Binder, D., & Naul, R. (2017). Other international efforts in the promotion of Olympic education. In R. Naul, et al. (Eds.), *Olympic education. An international review* (pp. 28–32). London and New York: Routledge.

Naul, R., Binder, D., Rychtecky, A., & Culpan, I. (2017). The promotion of Olympic education by the International Olympic Committee. In R. Naul, et al. (Eds.), *Olympic education. An international review* (pp. 16–27). London and New York: Routledge.

Naul, R., & Binder, D. (2017). Historical roots of the educational ideas of Pierre de Coubertin. In R. Naul, et al. (Eds.), *Olympic education. An international review* (pp. 9–15). London and New York: Routledge.

Utesch, T., Dreiskämper, D., Naul, R., & Geukes, K. (2017). The importance of accuracy in overweight and underweight children's perceived motor competence for future physical activity. In *NASPSPA 2017 Conference* (50th). San Diego, USA, (n.p.).

Naul, R. (2017). Pilot evaluation of the ASC Toolkits – Intermediate results. *Sport and Citizenship, 40*, 12.

Naul, R. (2017). Physical education and physical activity in active school communities. In A. Bund & C. Scheuer (Eds.), *Changes in childhood and adolescence: Current challenges for physical education* (p. 20). Berlin: Logos.

Naul, R., Dreiskämper, D., & Utesch, T. (2017). Web-based self-monitoring and evaluation of motor competence development of children with a daily combination of PE and PA in German and Dutch primary schools. In A. Bund & C. Scheuer (Eds.), *Changes in childhood and adolescence: Current challenges for physical education* (pp. 23–24). Berlin: Logos.

Onofre, M., Costa, J., Naul, R., Repond, R. M., Scheuer, C., & Holzweg, M. (2017). How to know more about physical education and school in Europe: A project for EUPEA. In A. Bund & C. Scheuer (Eds.), *Changes in childhood and adolescence: Current challenges for physical education* (pp. 44–46). Berlin: Logos.

Naul, R. (2017). Sport im Ganztag: Für wen und für was, womit und wozu? *SportPraxis, 58,* Sonderheft, 4–7.

Naul, R. (2017). CEREPS – der Europarat für Schulsport-Forschung und körperliche Aktivität. *FOSS- Newsletter 42*, 4.

Naul, R., & Binder, D. (2017). Olympic education: History, theory and practice. In D. Dumon, A. R. Hofmann, R. Diketmüller, K. Koenig, R. Bailey & C. Zinkler (Eds.),*Passionately Inclusive: Towards participation and friendship in sport* (pp. 159–174). Münster: Waxmann.

Naul, R. (2018). Die EU-Guidelines for physical activity. Entwicklung und Bedeutung. In J. Mittag (Ed.), *Europäische Sportpolitik* (pp. 271–286). Baden-Baden: Nomos.

Henning, L., Dreiskämper, D., Tietjens, M., Graf. C., Utesch, T., & Naul, R. (2018). Die Entwicklung des physischen Selbstkonzepts, der physischen Fitness und des BMIs im Kindesalter: Ein reziprokes Effektmodell.' In U. Borges, L. Bröker , S. Hoffmann, et al. (Eds.), *Abstractband der 50. Jahrestagung der asp* (pp. 87–88). Köln: DSHS.

Utesch, T., Dreiskämper, D., Naul, R., & Geukes, K. (2018). Understanding physical (in-) activity, overweight, and obesity in childhood: Effects of congruence between physical self-concept and motor competence. *Scientific Reports*, 8, 5908. doi:10.1038/s41598-018-24139-y

Utesch, T., Dreiskämper, D., Strauss, B., & Naul, R. (2018). The development of the physical fitness construct across childhood. *Scandinavian Journal of Medicine & Science in Sport, 28*(1), 212–219. doi:10.1111/sms.12889

Naul, R., Dreiskämper, D., Dorkewitz, I., L'Hoir, M., Heemsoth, S., Tietjens, M., & Strauss, B. (2018). Gesunde Kinder in gesunden Kommunen – Healthy children in sound communities (gkgk/HCSC). In WGI (Ed.), *WGI-Report 2013–2017* (pp. 32–39). Münster: WGI.

Naul, R. (2018). Olympische Erziehung – Olympic education In WGI (Ed.), *WGI-Report 2013–2017* (pp. 42–43.) Münster: WGI.

Scheuer, C., & Naul, R. (2018). The vision and mission of CEREPS, a European council of research in physical education and physical activity. In C. Scheuer, A. Bund, & M. Holzweg (Eds.), *Changes in childhood and adolescence. Current challenges for physical education* (pp. 64–66). Berlin: Logos.

Naul, R. (2018). Physical education and physical activity in active school communities. In C. Scheuer, A. Bund, & M. Holzweg (Eds.), *Changes in childhood and adolescence. Current challenges for physical education* (p. 67). Berlin: Logos.

Naul, R., Utesch, T., & Dreiskämper, D. (2018). Monitoring motor development: Measuring motor competence via an online database. In C. Scheuer, A. Bund, & M. Holzweg (Eds.), *Changes in childhood and adolescence. current challenges for physical education* (pp. 76–80). Berlin: Logos.

Onofre, M., Costa, J., Naul, R., Repond, R.-M., Scheuer, C., & Holzweg, M. (2018). How to know more about physical education and school report in Europe: EuPEO, a project from EUPEA. In C. Scheuer, A. Bund, & M. Holzweg (Eds.), *Changes in childhood and adolescence. Current challenges for physical education* (pp. 97–99). Berlin: Logos.

Naul, R. (2018). Identical global terminology with different continental meanings vs. different continental terminology with equivalent global meanings. In C. Scheuer, A. Bund, & M. Holzweg (Eds.), *Changes in childhood and adolescence. Current challenges for physical education* (pp. 169–184). Berlin: Logos.

Naul, R., Binder, D., Rychtecky,A., & Culpan, I. (Eds.). (2018). *Olympic education. An international review.* London and New York: Routledge (Paperback).

Naul, R., Utesch T., & Niehues, D. (2018). Linking physical education with local sport and organizations: Implementation of an "Active school community". *European Journal of Physical Education and Sport Science, 4*(11), 1–27. doi:10.5281/zenodo.1411519

Onofre, M., Costa, J., Naul, R., & Scheuer, C. (2018). Monitoring the quality of physical education: EuPEO, an ongoing project from the European Physical Education Association. In S. Gray & M. Thorburn (Eds.), *Book of abstracts. AIESEP World Congress 2018* (p. 199). Edinburgh: University of Edinburgh.

Naul, R. (2018). Willibald Gebhardt Institute. In M. Krüger & K. Reinhardt (Eds.), *Sport for all. History of a vision around the world. Book of abstracts of the 19th ISHPES Congress* (p. 21). Münster: WWU.

Naul, R. (2018). Different historical pathways – Different concepts of Olympic education in Germany. In M. Krüger & K. Reinhardt (Eds.), *Sport for all. History of a vision around the world. Book of abstracts of the 19th ISHPES Congress* (p. 99). Münster: WWU.

Naul, R. (2018). Session WGI: "History of Olympic education in Japan and Germany". In M. Krüger & K. Reinhart (Eds.), *Sport for all. History of a vision around the world. Book of abstracts of the 19th ISHPES Congress in Münster/Germany* (p. 97). Münster: WWU.

Dreiskämper, D., & Naul, R. (2018). Modul Bewegung. Gesundheitsförderung in der Ganztagsschule In M. Munser-Kiefer & E. Göttlein (Eds.), *Studienbuch Gesundheitsförderung.– Praxis für die Ganztagsschule* (pp. 14–29). Hiltpoltstein: afg media.

Naul, R., & Uhler-Derigs, H. G. (2019). *Sportunterricht in der Berufsbildung – Didaktisache Konzepte und Unterrichtsplanungen.* Aachen: Meyer & Meyer (Edition Schulsport, Band 40).

Naul, R. (2019). Bildungshistorische und bildungstheoretische Grundlagen: Die Förderung der körperlichen Erziehung im Kontext von Allgemeinbildung und Berufsbildung. In R. Naul & H. G. Uhler-Derigs (Eds.), *Sportunterricht in der Berufsbildung. Didaktische Konzepte und Unterrichtsplanungen* (pp. 13-48). Aachen: Meyer & Meyer.

Naul, R. (2019). Leibeserziehung und Sportunterricht in Berufsschulen/Berufskolleg in der Bundesrepublik Deutschland. In R. Naul & H. G. Uhler-Derigs (Eds.), *Sportunterricht in der*

Berufsbildung. Didaktische Konzepte und Unterrichtsplanungen (Edition Schulsport vol. 40) (pp. 49-129). Aachen: Meyer & Meyer.

Binder, D., Naul, R., & Fialova, L. (Eds.). (2019). Olympic education – History, theory, practice (Willibald Gebhardt Schriftenreihe, Band 16). Aachen: Meyer & Meyer (in print).

Naul, R., & Scheuer, C. (Eds.). (2019) *Research on physical education and school sport in Europe* (Edition Schulsport, Band 38). Aachen: Meyer & Meyer (in print).